THE HANDY ENGINEERING ANSWER BOOK

Visible Ink Press®
43311 Joy Rd., #414
Canton, MI 48187-2075

Most Visible Ink Press books are available at special quantity discounts when purchased in bulk by corporations, organizations, or groups. Customized printings, special imprints, messages, and excerpts can be produced to meet your needs. For more information, contact Special Markets Director, Visible Ink Press, www.visibleink.com, or 734-667-3211.

Managing Editor: Kevin S. Hile
Art Director: Mary Claire Krzewinski
Page Design: Cinelli Design
Typesetting: Marco Divita
Proofreaders: Larry Baker and Shoshana Hurwitz
Indexer: Larry Baker

Cover images: Shutterstock.

ISBN: 978-1-57859-770-3 (paperback)
ISBN: 978-1-57859-802-1 (hardcover)
ISBN: 978-1-57859-612-6 (ebook)

Cataloging-in-Publication Data is on file at the Library of Congress.

Printed in the United States of America

10 9 8 7 6 5 4 3 2 1

THE HANDY ENGINEERING ANSWER BOOK

DeLean Tolbert Smith, Ph.D.; Aishwary Pawar;
Nicole Pitterson, Ph.D.; and Debra-Ann C. Butler, Ph.D.

Detroit

Also from Visible Ink Press

The Handy Accounting Answer Book
by Amber Gray, Ph.D.
ISBN: 978-1-57859-675-1

The Handy African American History Answer Book
by Jessie Carnie Smith
ISBN: 978-1-57859-452-8

The Handy American Government Answer Book: How Washington, Politics, and Elections Work
by Gina Misiroglu
ISBN: 978-1-57859-639-3

The Handy American History Answer Book
by David L. Hudson Jr.
ISBN: 978-1-57859-471-9

The Handy Anatomy Answer Book, 2nd edition
by Patricia Barnes-Svarney and Thomas E. Svarney
ISBN: 978-1-57859-542-6

The Handy Answer Book for Kids (and Parents), 2nd edition
by Gina Misiroglu
ISBN: 978-1-57859-219-7

The Handy Armed Forces Answer Book
by Richard Estep
ISBN: 978-1-57859-743-7

The Handy Art History Answer Book
by Madelynn Dickerson
ISBN: 978-1-57859-417-7

The Handy Astronomy Answer Book, 3rd edition
by Charles Liu, Ph.D.
ISBN: 978-1-57859-419-1

The Handy Bible Answer Book
by Jennifer Rebecca Prince
ISBN: 978-1-57859-478-8

The Handy Biology Answer Book, 2nd edition
by Patricia Barnes Svarney and Thomas E. Svarney
ISBN: 978-1-57859-490-0

The Handy Boston Answer Book
by Samuel Willard Crompton
ISBN: 978-1-57859-593-8

The Handy California Answer Book
by Kevin S. Hile
ISBN: 978-1-57859-591-4

The Handy Chemistry Answer Book
by Ian C. Stewart and Justin P. Lamont
ISBN: 978-1-57859-374-3

The Handy Christianity Answer Book
by Steve Werner
ISBN: 978-1-57859-686-7

The Handy Civil War Answer Book
by Samuel Willard Crompton
ISBN: 978-1-57859-476-4

The Handy Communication Answer Book
by Lauren Sergy
ISBN: 978-1-57859-587-7

The Handy Diabetes Answer Book
by Patricia Barnes-Svarney and Thomas E. Svarney
ISBN: 978-1-57859-597-6

The Handy Dinosaur Answer Book, 2nd edition
by Patricia Barnes-Svarney and Thomas E. Svarney
ISBN: 978-1-57859-218-0

The Handy English Grammar Answer Book
by Christine A. Hult, Ph.D.
ISBN: 978-1-57859-520-4

The Handy Forensic Science Answer Book: Reading Clues at the Crime Scene, Crime Lab, and in Court
by Patricia Barnes-Svarney and Thomas E. Svarney
ISBN: 978-1-57859-621-8

The Handy Geography Answer Book, 3rd edition
by Paul A. Tucci
ISBN: 978-1-57859-576-1

The Handy Geology Answer Book
by Patricia Barnes-Svarney and Thomas E. Svarney
ISBN: 978-1-57859-156-5

The Handy History Answer Book: From the Stone Age to the Digital Age, 4th edition
by Stephen A. Werner, Ph.D.
ISBN: 978-1-57859-680-5

The Handy Hockey Answer Book
by Stan Fischler
ISBN: 978-1-57859-513-6

The Handy Investing Answer Book
by Paul A. Tucci
ISBN: 978-1-57859-486-3

The Handy Islam Answer Book
by John Renard, Ph.D.
ISBN: 978-1-57859-510-5

The Handy Law Answer Book
by David L. Hudson, Jr., J.D.
ISBN: 978-1-57859-217-3

The Handy Literature Answer Book
by Daniel S. Burt and Deborah G. Felder
ISBN: 978-1-57859-635-5

The Handy Math Answer Book, 2nd edition
by Patricia Barnes-Svarney and Thomas E. Svarney
ISBN: 978-1-57859-373-6

The Handy Military History Answer Book
by Samuel Willard Crompton
ISBN: 978-1-57859-509-9

The Handy Mythology Answer Book
by David A. Leeming, Ph.D.
ISBN: 978-1-57859-475-7

The Handy New York City Answer Book
by Chris Barsanti
ISBN: 978-1-57859-586-0

The Handy Nutrition Answer Book
by Patricia Barnes-Svarney and Thomas E. Svarney
ISBN: 978-1-57859-484-9

The Handy Ocean Answer Book
by Patricia Barnes-Svarney and Thomas E. Svarney
ISBN: 978-1-57859-063-6

The Handy Pennsylvania Answer Book
by Lawrence W. Baker
ISBN: 978-1-57859-610-2

The Handy Personal Finance Answer Book
by Paul A. Tucci
ISBN: 978-1-57859-322-4

The Handy Philosophy Answer Book
by Naomi Zack, Ph.D.
ISBN: 978-1-57859-226-5

The Handy Physics Answer Book, 3rd edition
By Charles Liu, Ph.D.
ISBN: 978-1-57859-695-9

The Handy Presidents Answer Book, 2nd edition
by David L. Hudson
ISB N: 978-1-57859-317-0

The Handy Psychology Answer Book, 2nd edition
by Lisa J. Cohen, Ph.D.
ISBN: 978-1-57859-508-2

The Handy Religion Answer Book, 2nd edition
by John Renard, Ph.D.
ISBN: 978-1-57859-379-8

The Handy Science Answer Book, 5th edition
by The Carnegie Library of Pittsburgh
ISBN: 978-1-57859-691-1

The Handy State-by-State Answer Book: Faces, Places, and Famous Dates for All Fifty States
by Samuel Willard Crompton
ISBN: 978-1-57859-565-5

The Handy Supreme Court Answer Book
by David L Hudson, Jr.
ISBN: 978-1-57859-196-1

The Handy Technology Answer Book
by Naomi E. Balaban and James Bobick
ISBN: 978-1-57859-563-1

The Handy Texas Answer Book
by James L. Haley
ISBN: 978-1-57859-634-8

The Handy Weather Answer Book, 2nd edition
by Kevin S. Hile
ISBN: 978-1-57859-221-0

The Handy Western Philosophy Answer Book: The Ancient Greek Influence on Modern Understanding
by Ed D'Angelo, Ph.D.
ISBN: 978-1-57859-556-3

The Handy Wisconsin Answer Book
by Terri Schlichenmeyer and Mark Meier
ISBN: 978-1-57859-661-4

Please visit the "Handy Answers" series website at www.handyanswers.com.

About the Authors

DeLean Tolbert Smith, Ph.D. is an assistant professor in the Industrial and Manufacturing Systems Engineering department at the University of Michigan-Dearborn. She earned her Ph.D. in Engineering Education from Purdue University. She investigates topics related to engineering design thinking and STEM informal learning. This research led to her earning an NSF CAREER award titled Investigating Black Youths' Engineering, Innovation, and Design Practices at the Intersection of Museum and Home/Family Learning. She is committed to creating resources that introduce engineering to many new audiences. She also has degrees in Electrical Engineering and Industrial Engineering. She and her husband live in Detroit, Michigan.

Aishwary Pawar is a doctoral candidate in Industrial & Systems Engineering at the University of Michigan-Dearborn. His main research interest centers on investigating the factors that influence undergraduate enrollment, retention, graduation, and dropout. For his master's thesis, Aishwary researched how student demographics and background characteristics lead to a more comprehensive understanding of a student's enrolment and retention at an undergraduate college. For his Ph.D. thesis, he is working under the supervision of Dr. DeLean Tolbert Smith. Currently, his research is focused on using human-centered design and data analytics to improve student access and success in an undergraduate engineering program and support higher education professionals to recognize minoritized students' diverse needs. Aishwary also works as a graduate student Instructor at the University of Michigan-Dearborn, where he teaches lab sessions in Engineering and Engineering Design.

Nicole P. Pitterson, Ph.D. is an assistant professor in the Department of Engineering Education at Virginia Tech. She holds a B.Sc. in Electrical and Electronic Engineering from the University of Technology, Jamaica, an M.Sc. in Manufacturing Engineering from Western Illinois University, and a Ph.D. in Engineering Education from Purdue University. Upon completion of her doctoral degree, she worked as a postdoctoral research scholar at Oregon State University. Her research interests include difficult concepts in engineering, increasing students' conceptual understanding of complex concepts, curriculum design, and promoting collaboration through using active learning strategies as well as exploring students' disciplinary identities through engagement with knowledge. Nicole is dedicated to bridging the gap between theoretical concepts with practical applications. She also aims to guide students to develop a critical understanding of core engineering concepts that goes beyond rote memorization so that they can adapt to the changing demands of a global workforce. She currently resides in Christiansburg, Virginia.

Debra Butler, Ph.D., received her bachelor of arts degree from the University of Miami and her Ph.D. in educational policy, planning, and leadership with a focus on higher education from William & Mary. Dr. Butler's career spans over 20 years in higher education, working in student services, academic affairs, and program administration. She currently resides in Michigan with her husband and two beautiful daughters.

Photo Sources

Jonathan Chen: p. 59.

Paul Clarke: p. 294.

Signe Dons: p. 334.

Finnish Museum of Photography: p. 316.

Lynn Gilbert: p. 274.

Digital Globe: p. 55.

Thorston Hartmann: p. 228.

Isaac Newton Institute: p. 98.

Javier Kohen: p. 52.

Library of Congress: p. 304.

J. Doug McLean: p. 311.

MikeRun (Wikicommons): p. 105.

Musée des Papeteries Canson et Montgolfier: p. 306.

MuslimHeritage.com: p. 96.

Theresa Knott Psarianos: p. 357.

Rama (Wikicommons): p. 282.

Science History Institute: p. 376.

Shutterstock: pp. 4, 8, 11, 14, 16, 19, 20, 24, 32, 34, 43, 46, 49, 51, 64, 70, 72, 73, 74, 75, 78, 81, 85, 87, 90, 93, 100, 101, 104, 107, 111 (edits by Kevin Hile), 113, 122, 132, 136, 137, 138, 141, 144, 146, 152, 155, 161, 167, 168, 172, 174, 180, 184, 189, 190, 194, 197, 199, 213, 216, 221, 224, 225, 232, 234, 235, 240, 243, 245, 248, 253, 254, 257, 259, 266, 269, 270, 271, 278, 280, 297, 300, 307, 315, 318, 320, 324, 337, 341, 343, 352, 353, 356, 361, 370, 375, 379, 382, 385, 387, 388, 399, 407.

Smithsonian Institution: p. 205 (right).

Daniel Stroud: p. 289.

Antoine Taveneaux: p. 83.

R. Thyroff: p. 176.

U.S. Food and Drug Administration: p. 218.

University of Manchester: p. 103.

Wapcaplet (Wikicommons): p. 115.

Yale University Art Gallery: p. 135.

Public Domain: pp. 10, 68, 170, 193, 203, 205 (left), 208, 286, 287, 335.

Table of Contents

Acknowledgments

This book is an attempt to convey engineering fundamentals to the reader with a particular goal of helping people from non-engineering backgrounds to understand the engineering culture, disciplines, domains, its challenges, and vast opportunities. The authors placed a strong emphasis on balancing historically and culturally relevant facts with practical engineering knowledge. This involved a tremendous amount of effort and the crucial contributions of several insightful and encouraging individuals. We want to acknowledge those who advised and provided editorial support on this work: Drs. Tamecia Jones, Oluwaseyi Ogebule, Dorian Davis, and Ms. Rita Brooks.

A personal note from DeLean Tolbert Smith: I am grateful for the opportunity to communicate my love of engineering to a new community of readers. Many of you I may never meet, but I hope that this book serves as a fantastic starting point or reference guide along your engineering journey. I would like to acknowledge my family for their support—specifically, my mother, Myrtle, and her constant pursuit engineering excellence and my husband, Terrence, for his unwavering support.

A personal note from Aishwary Pawar: First and foremost, I'd like to express my gratitude to Dr. DeLean Tolbert Smith, my adviser and co-author, for including me in this project and for the continuous support and patience throughout this work. I owe an enormous debt of appreciation to Roger Jänecke, the publisher, for his detailed and constructive feedback. An additional thanks to the University of Michigan-Dearborn for giving me a platform to teach, train, learn, and improve. Iíd like to acknowledge the assistance of *The Handy Engineering* editorial team for the invaluable editorial assistance. Finally, I am extremely grateful to my parents for their love, support, and unending inspiration.

Introduction

Engineering impacts every aspect of our lives. Throughout history, engineering ideas and innovative feats have provided solutions that still confound modern historians, scientists, and engineers. In this book, we sought to provide a resource that both engineers and non-engineers alike could reference to learn fundamental facts about engineering. We hope that *The Handy Engineering Answer Book* not only provides useful information but also inspires more people to consider studying engineering and working professionally as engineers.

Engineering is about identifying a need and developing solutions for that need. That is why we wrote this book. We saw an opportunity to provide a book for readers who have a strong interest in the field, for those who are exploring engineering for the first time, and for those who fall in between.

We hope that you see that engineering is much more than providing one solution in a specific discipline and using one specific process or approach. Rather, engineers work to understand problems and needs fully; they offer out-of-the-box ideas, use science and technology and mathematics to develop solutions that work well, build functional prototypes for testing, and get feedback from the user for improvement!

The Handy Engineering Answer Book is organized in a way that is easy to follow:

Chapter 1: Introduction to Engineering and History answers questions related to the history of engineering and includes a timeline of historical happenings that we believe everyone should know.

Chapters 2 through 11: Engineering Disciplines provide a general overview of the major divisions within engineering and their respective subdisciplines. In addition to the overview, we also shared fun facts, bits of history, and details about important people in each discipline.

We wrap up with Chapter 12: Engineering Pathways, which is important to provide resources that could be helpful for readers considering engineering at every level. This chapter provides answers about organizations, college expectations, and ways to get involved in engineering academically, professionally, or for fun. An appendix at the end of the book provides additional resources for further education and on important engineering organizations.

Introduction to Engineering

What is engineering?

Engineering is a branch of science that deals with the application of mathematical and scientific principles. It is also a discipline and profession that meets the needs of society. Engineers incorporate interpersonal skills such as teamwork, creativity, decision making, ethics, leadership, and design to create solutions to societal problems and improve daily life. People who work in engineering analysis generate and gather data that they use to think critically, make decisions, and create new knowledge. To accomplish these goals, engineers need a strong background in science and mathematics.

Engineers work in many different types of work settings. They can be found wearing hardhats on a construction site, on a farm improving agricultural technology such as dams or water reservoirs or developing biofuels from food. Engineers can also be found in fancy offices while helping businesses develop models that can predict their future success.

Where does the word engineer come from?

The word "engineer" is derived from the Latin words *ingeniare* and *ingenium. Ingeniare* means to create, generate, contrive, and devise, and *ingenium* means cleverness.

What is the goal of engineering?

The primary goal of engineering is to make people's lives better. In this regard, engineers seek to identify problems with practical significance to local and broader contexts. Through their work, engineers endeavor to make technological advances using innovative techniques along with their application of mathematical and scientific knowledge. The goals of engineering can be broken down into two main categories: creating processes and designs and maintenance and operations.

Engineering processes and designs: Engineers create and use processes to perform complicated tasks in a fairly easy manner. Oftentimes, in engineering work, the engineer will devote a significant amount of time at the start of a project to ensure they have all the right tools, materials, and equipment to complete the task at hand. Then, they design the process by which the problem will be solved. Using their technical skills and knowledge, the engineer then designs, creates, and implements the solution. The process used to move from defining the problem to creating a solution is known as the engineering design process. This will be discussed in more detail later on.

Engineering maintenance and operations: Engineers also use their technical skills and knowledge to maintain systems to ensure that they work as intended. Sometimes, it is necessary for the engineer to troubleshoot and repair a certain portion of the system or to teach others how to use technologies within the workplace. Engineering work is governed by professional bodies and government regulations and, as such, engineers must ensure that where they work and what they do are in accordance with these rules and regulations as well as all ethical and safety procedures.

What is the difference between science and engineering?

Science is defined as the body of knowledge that is used to investigate and discover new things about the natural world. Science helps us understand how nature works, while engineering applies scientific knowledge to make sense of the world and solve problems. Scientists and engineers use some of the same fundamental knowledge but contribute to the world in different ways.

What is the difference between science and engineering *processes*?

Scientists use the scientific method to discover or understand natural phenomena. Scientists begin by asking questions, gathering background information, making some assumptions and hypotheses, and testing these assumptions and hypotheses by conducting experiments to collect and analyze data, which is then reported as their findings.

Engineers use the design process to create solutions. They tend to start with defining the problem, determining the criteria and constraints associated with that problem from which they brainstorm ideas, devising a plan of action, and making a prototype that is then tested and modified before being implemented.

What are examples of engineering tools?

Beyond combining the principles of mathematics and science, engineering embodies other practical areas such as technology, economics and commerce, communication, design, statistics, and innovation. The tools of engineering often encompass an integration of knowledge and skills aimed at creating optimal solutions with the least amount of risk and harm. A single right answer does not exist in engineering. Instead, engineering teams often make decisions that take into consideration constraints, available resources, time required from idea generation to completion, preferred end goal, and most efficient way to arrive at the goal.

Who are engineers?

Engineers are individuals who are possessed with and demonstrate fundamental, discipline-specific knowledge, outstanding problem-solving skills,

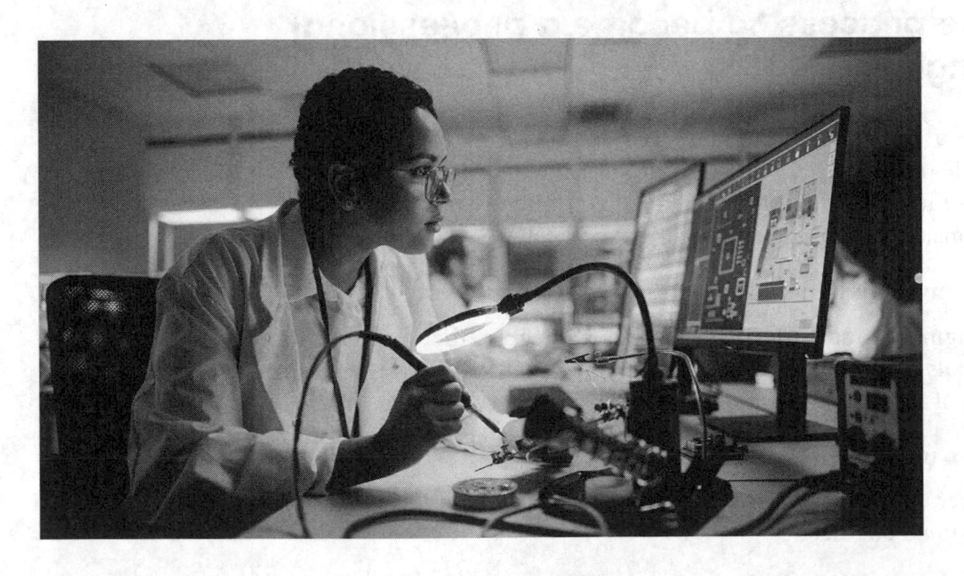

Engineers are people who use scientific and mechanical knowledge to solve problems by designing machines, tools, and devices and then turning them into reality.

and highly ethical decision-making. As practitioners, engineers embody traits such as providing service with the goal of addressing human and social needs, participating in professional practice such as engineering work teams and other professional communities, and engaging in critical, design, and reflective thinking in various contexts.

Is engineering an official profession?

Yes; all engineering disciplines and jobs require a bachelor's degree in engineering. However, some disciplines, like civil engineering, require that the engineer earn their bachelor's degree first, spend years working in the field, then take a Principles and Practice of Engineering (PE) exam in order to get a license. Being a professional engineer means different things to different people. For example, to a client, being a professional engineer means that the engineer has the required qualifications to perform the job at hand, while to an employer, being a professional engineer means that the engineer can perform at a higher level than someone who just graduated from college with an engineering degree. A professional engineer is able to demonstrate high levels of intellectual ability as well as the practical know-how to create solutions that serve the greater good. They are expected to use their extensive knowledge, experience, and skills to offer engineering solutions and services to the public. The laws of licensing vary from state to state in the United States; however, no matter where they are or where they work, all PEs are bound by a code of ethics and should create products and designs that protect the health and safety of the communities they serve.

What is the process to become a professional licensed engineer?

To become a professional engineer, the engineering graduate, having completed their degree at an accredited institution, must then work under a PE for a period of not less than four years, sit and subsequently pass two competency examinations, then earn their state-approved license.

Step 1: Become an intern

After an engineer graduates with a bachelor's degree, they will take the Fundamentals of Engineering (FE) exam and earn the title Engineering Intern (EI) or Engineer in Training (EIT).

Step 2: Gain professional experience

The engineer will then work for at least four years under the supervision of a professional engineer.

Step 3: Learn state level licensure requirements and qualifications

While the engineer is gaining professional experience, they will seek out their state's exam and qualifications requirements. Each U.S. state has its own exam and required qualifications.

Step 4: Prepare for and take the PE exam

Once the engineer is ready, they will prepare to get their license by studying for and completing the Principles and Practice of Engineering (PE) exam.

After earning their PE license, an engineer is required to maintain this license so as to continually maintain and improve their technical knowledge and skills. The PE must always be aware of and complete several state-regulated, continuing-education requirements so as to keep their license current.

What do engineers do?

Engineers work in a variety of contexts using their knowledge of mathematics and science to develop safe, economically sound, and context-specific solutions to everyday problems. These problems might be of various scales—small, medium, or large—and complexity. Engineering work is not always about creating new solutions, as sometimes, engineers work to improve and maintain existing systems and/or processes.

Engineering practice falls into three broad categories:

- Problem solving incorporates the systematic process that engineers use to scope, define, and solve problems of varying natures.
- The knowledge specific to each discipline that is necessary to engage in the problem-solving process.
- The integration of engineering problem-solving processes and knowledge.

What are the different types of engineers?

Many different types of engineering disciplines and engineers exist. Each discipline has its own body of knowledge, set of required skills, and subdisciplines. Each chapter of this book is dedicated to each of the major disciplines.

- Aerospace engineers: deal with the creation, design, and production of aircraft (for commercial, military, or private use) and other space exploration vehicles. Other career paths associated with aerospace engineering include systems engineers, aeronautical engineers as well as engineering project managers.

- Agricultural and biological engineers: broadly encompass designing and analyzing systems for the use and protection of natural resources, support for agricultural and biological systems, as well as providing environmental and climate controls. Agricultural and biological engineers are often exposed to the breadth of the agricultural industry including farming machinery, economics, resource management and the development of sustainable systems, livestock and food farming etc.

- Automotive engineers: using technology, automotive engineers research, design, and develop vehicles that adhere to the highest level of safety systems and standards. In some cases, their jobs also include the manufacture of vehicles and their subsystems.

- Biomedical engineers: combine discipline-specific knowledge associated with medicine, biology, and engineering. Biomedical engineers design, engage in, and produce systems that are directly related to patient care. These engineers are also closely associated with the development of healthcare innovations as well as sophisticated medical treatment and products.

- Chemical engineers: use the principles of chemistry found all around us in the form of atoms and matter to create products to meet specific needs. Using the molecular structure of materials and substances, chemical engineers are able to apply chemical systems and processes to technology to produce plastics, paint, fuel, medicines, fertilizer, semiconductors, paper, among various other products.

- Civil engineers: specialize in various areas such as transportation, structures such as roads, bridges etc., water resources and supply systems, building design, and construction. Civil engineering is one of the oldest engineering disciplines.

- Computer engineers: work primarily to use technology to continuously design, develop, and improve software and hardware and other computer-controlled equipment and processes. Computer engineers are often focused on improving their programming, troubleshooting, and maintenance skills as their work is so closely linked to advances in technology.

- Electrical engineers: tend to specialize in circuit analysis, power systems generation, transmission and distribution, communication, instrumentation, and controls. These engineers often design power systems and continuously work to improve the function of electrical equipment. Often considered to be the largest field of engineering because of its vast reach, electrical engineering operates on a macro (power grid for an entire country) and microscale (small electric devices such as a pacemaker).

- Environmental engineers: focus on ways to protect and improve the environment. Environmental engineering combines engineering and scientific principles to study the quality of air, water, and soil. They seek ways to manage and reduce pollution, minimize human impact on the planet as well as how to preserve the natural environment.

- Geological and geophysical engineers: study earth sciences and how changes in the makeup of the soil and other natural resource deposits can affect the foundation of buildings, roads, dams, and other infrastructure.
- Industrial engineers: embody the principle of efficiency and the most effective ways to use resources, e.g. people, materials, knowledge, money, energy etc. to achieve the best outcome with the least amount of waste.
- Manufacturing engineers: are often closely linked with industrial engineers. Manufacturing engineers often supervise and coordinate manufacturing processes and systems. From the product planning to the acquiring of material and creation of the final product, manufacturing engineers control automated systems, project design, and management.
- Marine and ocean engineers: design, produce, maintain, and improve commercial, military, and private ocean vehicles, marine propulsion structures, and systems. This set of engineers manage ocean transportation systems and vehicles aimed at exploring and managing natural resources.
- Mechanical engineers: study the influence of force, energy, and motion. Mechanical engineers work on a breadth of projects that often include other engineering disciplines such as aerospace, electrical, manufacturing, computer, biomedical, etc.
- Mining engineers: explore various methods of extracting mineral deposits from the earth. These engineers design and use extraction equipment and processes, and they supervise the creation of mines, the extraction process, and the subsequent closure of the mine. The process of mining has negative effects on the environment. Therefore, mining engineers also have to consider ways to minimize the harsh impact of mining.
- Nuclear engineers: study the use of nuclear energy, harnessed from atoms, to create and sustain power systems. Some nuclear engineers work for the military as well as other commercial enterprises. The impact of nuclear engineers is also manifested in the industrial and health sectors where radiation is used in cancer treatments as well as other medical diagnostic machines and processes.
- Petroleum engineers: develop and design systems to discover and extract oil and gas from reservoirs in the earth. These engineers not only design the oil wells and extraction system, they also determine the safest way to transport the oil and natural gas from the point of extraction to where they are stored and processed. Additionally, petroleum engineers explore how to extract oil and gas from older wells in order to meet the demand for these resources across the world.

What are greatest engineering achievements of the twentieth century?

In engineering it is important to understand the accomplishments of the past so that we can advance the work and impact of these technologies. Nearly everything around us that makes our lives easier was designed and/or created by engineers. From being able to turn on the faucet to getting clean drinking water, preparing food, getting from one place to another via automobiles, and communicating with family and friends, engineers in our society have significantly changed the way we live and exist since the beginning of civilization. These are some of the many areas in which engineers were key:

1. Electrification
2. Automobile manufacturing
3. Airplane design and manufacture
4. Water supply and distribution
5. Electronics
6. Radio and television
7. Mechanization of agriculture
8. Computer design and manufacture
9. Telephones and telecommunication
10. Air conditioning and refrigeration
11. Highways and other road structures
12. Spacecraft and space travel
13. Internet
14. Imaging

Engineers are involved in almost every aspect of our modern lives. One obvious place you'll see them working is in our auto industry.

15. Household appliances
16. Healthcare technologies and advances in medicine
17. Petroleum and petrochemical technologies
18. Laser and fiber optics
19. Nuclear technologies
20. High-performance materials

Who was considered one of the first great engineers?

Archimedes (287 B.C.E.–212 B.C.E.) was a Greek engineer, inventor, mathematician, and scientist whose work left a long-lasting impact on the fundamental theories that engineers use today. For example, the Archimedes Principle states that when an object is partially or fully immersed in a liquid, the object experiences a loss in its weight that is the same as the weight of the fluid that is displaced by the part of the object that is immersed. This is the general principal of buoyancy, and engineers need this information as they design floating vessels. Archimedes also designed the screw pump and specialized catapults and other weapons used in war. Many engineers were inspired by his contributions.

Who are some famous engineers from the 1800s through the early 1900s?

Many people have earned engineering degrees and used their knowledge and expertise to create many inventions that we use daily. Here are some well-known engineers:

It was German engineer Nikolaus Otto who was responsible for envisioning the compressed charge internal combustion engine that we still use today.

- Alexander Graham Bell (1847–1922): known for the creation of the first telephone. He also made significant contributions to the development of aeronautics and optical telecommunications.

- Thomas Edison (1847–1931): The incandescent lightbulb, the motion-picture camera, and the phonograph are among Edison's inventions. He is credited for founding the first electric power distribution company in New York City, providing 110 volts directly to 59 consumers using his newly created light bulbs, which he invented in 1879. In 1876, he also established the first industrial research laboratory.

- Henry Ford (1863–1947): known for his founding of the Ford Motor Company, which was one of the first companies to mass-produce vehicle parts and high-volume assembly lines, thereby creating cars that were affordable to most people.

- William S. Harley (1880–1943): an American mechanical engineer who was one of the founders of the Harley-Davidson Motor company.

- Nikolaus Otto (1832–1891): known for his development of the four-stroke engine, which has revolutionized the creation of motor vehicles. He developed the principle under which combustion engines are designed and produced.

- George Stephenson (1781–1948): an English civil and mechanical engineer known for the development of the first steam locomotive, which was used on the first railway system. In some texts, he is called the Father of the Railway.

- Nikola Tesla (1856–1943): Tesla is most recognized for inventing the alternating-current (AC) electrical system, which is currently the most widely used electrical system in the world. In addition, he invented the Tesla coil, which is still utilized in radio technology today.

- Wilbur (1867–1912) and Orville (1871–1948) Wright (aka the Wright brothers): known for their development of the airplane, these brothers were first bike mechanics who developed their love of flying kites into airplanes, which significantly changed the modern world.

Many other famous, modern-day engineers continue to create products and systems or have achieved life-changing feats, such as space travel, that will impact our lives and those of many generations to come. Some of these engineers are:

- William Sanford Nye, who is commonly known as Bill Nye the Science Guy (1955–), is an American mechanical engineer and science television personality.

- Jeff Bezos (1964–) is an American computer engineer and entrepreneur who is the founder of Amazon.

Bill Nye the Science Guy is a well-known popularizer of science for general audiences. He is also a mechanical engineer by training.

- Dr. Donna Auguste (1958–) is an African American electrical and computer engineer and entreprenuer who received a patent for the Personal Digital Assistant (PDA) while working for Apple Computer.
- Buzz Aldrin (1930–) is an American engineer, former astronaut, and fighter pilot who made three spacewalks. He and Neil Armstrong were the first two people to land and walk on the moon.
- Tom Scholz (1947–) is an American rock musician and mechanical engineer.
- Dr. Bonnie Dunbar (1949–) is an American engineer with degrees in ceramic, mechanical, and biomedical engineering. She is a former astronaut who flew five missions for NASA, logging over 50 days in space. Her final space flight was in 1998 to to the Mir space station.
- Dr. Marian R. Croak (1955–) is an African American engineer with over 200 patents to her name. Her most recognized invention is Voice Over Internet Protocols (VoIP). She is the vice president of engineering at Google.

What are some historic engineering events from around the world?

A man known as Imotep is believed to have been the first engineer. One of his most famous creations is the Step Pyramid in Saqqarah, Egypt, deter-

Year	Engineering Event
Neolithic Period 10000–3000 B.C.E.	• Methods to produce fire at will • Agricultural methods and machines • Artisans learn to melt materials to create tools and make bricks • Invention of textiles such as felt
The Copper Age 5000–3000 B.C.E.	• Development of the wheel and axle • Development of the plow • Use of stone tools • Written communication • Use of soft metals for tools • Babylonian Engineering Period 3000–600 B.C.E. • Familiarity with basic math • Development of the ability to calculate the area and volume of excavations • Development of the number system using base 60 • Primitive arches used in moving water • Bridges built with stone piers carrying wooden stringers • Roads built • Gardens of Babylon constructed
Egyptian Engineering Period 2900–1900 B.C.E.	• Pyramid Age • Development of the ability to calculate the size of stones
Roman Engineering Period 600 B.C.E.–400 C.E.	• Development of aqueducts for water supplies and sanitary systems
The Middle Ages 0–1500 C.E.	• Development of the first printing press • Leonardo da Vinci: architect, engineer, and artist • Development of feats for war, such as bridges and catapults
The Industrial Revolution 1600–1700	• James Watt built the steam engine • Development of spinning and weaving machinery

Year	Engineering Event
The Industrial Revolution 1600–1700 (contd)	• Founding of Luigi Galvani's principles of electrical conduction
The Revival of Science 1700–1800	• Robert Hooke discovered the elastic limit • Christiaan Huygens discovered the spiral watch spring and the pendulum clock • Sir Isaac Newton developed the laws of motion and calculus
Modern Science Begins 1800–1899	• Further development of electricity • Evolution of electricity-generation methods • Transmission of electrical signals • Development of iron-refinement methods
Twentieth-century Technology	• Henry Ford built cars • Thomas Edison developed electrical equipment • The Wright brothers designed and created the first airplane • Development of nylon and plastics • Creation of the first computer • Creation of transistors using silicon • Creation of jet engines • Development of laser technologies • Evolution of satellite communications
Twenty-first Century Engineering	• Construction and opening of the Burj Khalifa, the world's tallest building • Advent of Bluetooth technology • Creation of robotic body parts, such as arms or legs • Construction and opening of the Millau Viaduct, the world's tallest bridge • Opening of the Large Hadron Collider, an advanced research facility, to test the Higgs boson particle theory

The Burg Khalifa in Dubai, UAE, is currently the world's tallest building, standing at 2,722 feet (829.4 meters) tall.

mined to have been built in 2550 B.C.E. Since then, we have seen significant success across many regions of the world.

The table on pages 12–13 shows a timeline of engineering events throughout history.

What does it mean to be an engineer?

Over the course of history and even now, engineering knowledge evolved to meet the needs of the society in which it exists. In the nineteenth century, engineering culture was centered on the idea of apprenticeship. In an apprenticeship, a student learns the skills of a trade from senior personnel. In the United States, most engineers served as apprentices in the shop or field under the mentoring of senior engineers. Upon completion of their training, they were considered engineers themselves, possessing the relevant skills and knowledge to perform tasks associated with their job requirements competently.

At the same time, in other cultures, preparation for engineering work encompassed exposure to industrial, commercial, and other publicly focused schooling activities. As the centuries changed and World Wars I and II were fought, the goal of an engineering education alternated between focusing on practical, hands-on training and knowledge of purely scientific concepts.

Does the definition of what an engineer does vary from one country or region to another?

- The United Kingdom: In the nineteenth century, Isambard Kingdom Brunel, a famous engineer, gained fame through his design of chains for the SS *Great Eastern* steamship. Out of his legacy and that of similar engineers at the time, a misunderstanding grew that the word engineer referred to people with a high school education and expert skills in a certain trade. This included people who were knowledgeable about repairing small home appliances, cars, or refrigerators or even those who worked at power stations, developed infrastructure, or had other complex roles such as aircraft design and other complex systems. Instead, the definition of what counts as engineering among educators encompassed people with a university-level education and training who completed certain external qualifications and competency assessments.

- Europe: Across the continent, several countries have adopted the definition of engineer as someone who is highly skilled, educated, and certified by the European Federation of National Engineering Associations (FEANI). FEANI is an external board that assesses the engineers' experience and competence in applying scientific knowledge, professional skills, and safe and ethical principles as well as responsible behavior toward the environment and society. Once certified, the engineer can use the title EUR ING on all of their identifying documents such as passports and driver's licenses.
- United States: In the 1800s, an engineering education was promoted as a way of applying science learning to improving life. Derived mostly from the French school of thought, engineering in the United States meant undergoing training from a professional school of engineering, where the goal was to teach hands-on skills so that the country had sufficient manpower who were scientifically and technically well trained. Since then, what it means to be an engineer has not changed much. To practice as a professional engineer, one must be licensed through their home state (see the section about professional engineers later). The professional practice of engineering is governed by the National Society of Professional Engineers, and all engineering schools must be accredited by the Accreditation Board for Engineering and Technology (ABET).
- Canada: Based on its European roots in the early centuries, being an engineer in Canada meant working in the military to build weapons of war or to devise systems to give the army an advantage. In the nineteenth century, the practice and profession of engineering took on a nonmilitary perception and instead leaned more toward civil engineering. Over time, more specialization emerged as society's needs for technological and infrastructural development increased. Today, the profession of engineering ranks as the third most recognized profession. Professional engineering practice is governed by the Canadian Council of Professional Engineers.
- China: Chinese scientists and engineers have made significant contributions to the world we live in today. Some of the earliest advancements are centered around science, technology, engineering, mathematics, and

astronomy. Early inventions that have been attributed to China include the ability to observe comets and solar eclipses, the creation of the abacus, and flying devices such as kites and Kongming lanterns. Four great inventions occurred in China in the early sixteenth and seventeenth centuries. These were the compass, gunpowder, papermaking, and printing. China also made some of the oldest contributions to medicine, such as acupuncture and herbal treatments, as well as to the disciplines of mechanical engineering, architecture, and civil engineering. Today, China's influence on engineering and science cannot be understated. In fact, China graduates more engineers on a yearly basis than the United States, Japan, and Germany combined, and they have been reported as having the highest science output, which is measured by the length and breadth of studies published in any given year. Being an engineer in China is highly prestigious, as this profession is considered the most influential and critical for nationwide mobility.

- Africa: This continent is known for and has been reported as having the oldest and most technologically advanced achievement in human development. The Great Pyramids are testament to this fact! Africa has been credited with the development of several key aspects of what engineering is and what it means to be an engineer, such as mathematics, astronomy, metallurgy and tools, architecture, medicine, and navigation among many other applicable concepts. Because of the vast size and number of nations (52) in this continent, what is perceived as engineering varies by region. Africa is home to some of the world's oldest educational institutes and facilities, such as the Library of Alexandria (295 B.C.E.), Al-Azhar University (f. 970), and Cairo University among others. An example of a current engineering contribution includes AfroAgEn, which is an organization that uses engineering to develop solutions to

Of course, the Great Pyramids of Egypt are one of the greatest engineering feats of all time. Even today, modern engineers are unsure how these enormous structures were built by a society that, as far as we know, had only the simplest tools.

DID YOU KNOW?

What is the history of women in engineering?

Women have been earning engineering degrees since at least the 1890s and performing engineering work for hundreds of years before then. Unfortunately, women were routinely denied opportunities to earn graduate degrees, and few women were allowed admission into undergraduate engineering programs. In the United States, World War II caused a shift in the inclusion of women. Many men were sent off to fight in the war, and this left a large shortage of people who were trained to do engineering work such as repair large and small electronics. The General Electric company began an on-the-job training program for women who already had mathematics and science degrees. The Curtiss-Wright Corporation was an airplane-manufacturing company that created a program that prepared women to take jobs in the defense industry. Throughout history, women have also contributed to engineering inventions and improved processes and left their mark on society even if they did not have engineering degrees. Currently, women make up 20 percent of the undergraduate degrees earned in the United States each year in engineering. Some famous women engineers include Alice Perry (1885–1969), Cécile Butticaz (1884–1966), Ada Lovelace (1815–1852), Elizabeth Bragg (1858–1929), Aprille Ericsson-Jackson (1963–), Mary Jackson (1921–2005), and Mae C. Jemison (1956–).

food security and economic prosperity. Recently, Noël N'guessan of KubeKo (b. 1990?) was awarded the Africa Prize for Innovation for developing biowaste-processing equipment for farmers. In 2016, Peter Mbira (b. 1990) was awarded the same prize for his 4×4 off-road, multipurpose wheelchair design.

Today, engineering has even further evolved to include efforts focused on research, information literacy, design, and work with other professions and areas of learning. Engineers populate industry, factions of government, business, nonprofit organizations, and academia among many other professional divisions in society. Broadly, engineers work within specific contexts, taking into consideration the ethical, environmental, economic, and social impact of their work on the public.

Fun facts:

1. In Europe, the United States, and Canada, it is illegal to use the title "engineer" to offer services to the public if one is not licensed in accordance with the respective governing bodies of the country and/or state.

2. In some countries, such as Canada, Italy, and Spain, the engineering profession is highly regarded and sometimes even ranked above other professions. For example, in Italy, engineers can include their title with their names as one would use Miss, Mr., or Dr.

ENGINEERING DESIGN BASICS

What are the ways that engineers solve problems?

Engineers face problems that are considered "open-ended" problems. This means that the problems have more than one solution and that not every solution is the best solution. This is different from science and mathematics problem solving, which often provide the problem solver with enough details to solve for one correct solution. Open-ended problems can be complex and lack needed information that would allow the engineer to simply select the correct equation or method to solve the problem. The engineering method requires the engineer to use a different set of skills, which includes:

- Representing and communicating a design problem
- Making reasonable assumptions and understanding constraints
- Generating possible ideas
- Gathering information and searching for existing solutions
- Developing a plan of activities
- Using available resources well
- Using project management to organize project elements, activities, and team assignments

What are some examples of early engineering problems?

The word engineer is derived from the Latin words *ingeniare* and *ingenium*. *Ingeniare* means to create, generate, contrive, and devise, and *ingenium* means cleverness. Ancient people developed sophisticated solutions that still amaze engineers, scientists, and mathematicians today. One example of an early engineering problem was how to lift a weight that was heavier than a person can carry. A few examples of amazing structures throughout time demonstrate engineering feats. For example, ancient Egyptians and Mexicans solved this problem in order to build pyramids. Some engineering solutions were developed

The Leshan Giant Buddha was built around the eighth century during China's Tang Dynasty. Carved out of red sandstone where the Min and Dadu rivers meet, it stands 233 feet (71 meters) tall).

before the scientific theories that explained how they worked! Some examples of ancient engineering innovations are:

- Great Wall of China
- Saint Basil's Cathedral, Russia
- Taj Mahal, India
- Horyuji Temple, Japan
- Chand Baori, India
- Leshan Giant Buddha, China
- Borobudur, Indonesia
- Mohenjo-Daro, Pakistan
- Pyramid of Giza, Egypt
- Blue Mosque, Turkey
- Great Sphinx, Egypt
- Temple of Bacchus, Lebanon
- Great Mosque of Damascus, Syria

How do modern engineers use scientific theories?

Modern engineers use theories to develop models. Engineering models are helpful tools for representing and explaining how a theory works in a system, product, or process and will behave under certain conditions. Some models are physical prototypes; other models are computer simulations, diagrams, drawings, storyboards, or analogies and are constructed using mathematics. Models can be complex, or they can be more basic and simple. This is a core skill set for engineers! They must be able to understand scientific theories and laws and develop models that use their understanding to explain the world and natural phenomena and develop solutions. Some examples of engineers using scientific discoveries related to human vision are:

- Engineers used the scientific study of optics (the science of light interacting with matter) to engineer eyeglasses
- Chemists discovered improved materials for eyeglass lenses, and engineers used these discoveries to design thinner lenses and flight lenses
- Scientists improved contact lenses to be more water absorbent and softer
- Telescopes helped people see things that are far away and used scientific discoveries related to lenses
- Microscope designs used science discoveries related to light and the shape of lenses and helped to magnify tiny materials (making them look bigger than they actually are)

How do engineers and scientists work together?

Engineers and scientists both work diligently to ask questions about the world around them and problems they are faced with. The Hubble Space Telescope is the first telescope that was designed to allow astronauts to service it

Launched in 1990, the Hubble Telescope is still in operation and has been taking valuable photographs of distant objects in ultraviolet, near-infrared, and visible light spectrums ever since.

while in space (in-service). As of 2022, astronauts had visited the telescope five different times to fix problems or replace old parts.

How did scientists and engineers work together to service the Hubble Space Telescope?

In the first servicing mission to fix Hubble in 1993, scientists and engineers worked together to discover that the Hubble's main mirror had a defect and that the images it sent back to NASA were not clear. The astronauts installed a set of specialized lenses to correct its flawed main mirror. Scientists and engineers worked as a team on Earth to construct models and test them many times. They generated lots of data and used the analyzed data to make decisions about which of their solutions would solve the problem.

How do engineers use mathematics to solve problems with design?

Mathematics can be used in engineering to model real-world behavior. Engineers use the data that mathematics gives them to gather more information and make designs that create solutions. In fact, mathematics also has a role in design itself. For example, with the Hubble Space Telescope, mathematics and measuring were used to design the shape of the main mirror. Unfortunately, in that case, the mirror was incorrectly shaped by a small fraction, which caused the lens to capture blurred images. Similar to eyeglasses that help correct vision, engineers had to use their understanding of science and evidence from science experiments to create a set of devices that could correct the Hubble Space Telescope's blurred vision.

What is the importance of engineering design?

Engineering design is an interactive process that is used to identify and solve problems and to find the best solutions for a given scenario. This process is a very useful tool that is usually performed by a team of people. Everyone on the team will bring specific knowledge, skills, and experiences to the process. When a team has more diversity in knowledge, skills, and experiences, they typically develop better solutions. However, having a great team does not mean that all the solutions will always work. One characteristic of engineering design is that failure is expected. Failure actually helps teams find the best solutions before putting the final solution into practice. When an idea fails, the team has new information about the problem, solutions, and their ideas.

DID YOU KNOW?

How do engineers know that the product or artifact is acceptable?

When engineers are developing the design for a new product, they must consider its purpose, its form, and the environments in which it will be manufactured and operating. The purpose of the product can be determined by understanding the problem or the need that the product will address. The form of the product must be appropriate for its purpose. The engineer can determine if the form of the product is appropriate to the purpose by examining the operating environment (where it will be used) and the engineering environment (where it will be made). The operating environment can include natural factors (temperature, humidity, gravity), ways that people will interact with the product, and how other products interact with the form of the product. The engineering environment considers the form of a design with consideration to cost of manufacture, resources available, tools, and facilities. Some forms of a product may be acceptable in some environments and unacceptable in others. For example, consider different designs developed for the purpose of telling time (i.e., a sundial, a wristwatch, and a radio clock). The sundial will only work in a sunny operating environment; otherwise, it will not be useful.

Is the engineering design process used in only one way?

The engineering design process is extremely flexible and can be used on both engineering and nonengineering projects. The engineering design process can be approached in multiple acceptable ways. In fact, people have developed models to describe the process, the steps, and its iterative nature.

How does an engineer determine the performance of an acceptable design?

Engineers must have a basis to compare how well an acceptable design meets its purpose. They will test their design against the ideal performance measures and make necessary adjustments to make sure the product meets its goals. The engineer is constantly improving the product's form. In order for a product to be acceptable, it must meet cost and performance goals, which are determined by the operating and engineering environments.

What are the series of steps in the engineering design process?

The engineering design process does not have one single definition or description. Each version of the design process has some core elements, including:

- Recognizing the need for a product or service
- Defining the problem
- Developing possible solutions
- Creating prototypes
- Improving designs
- Communicating designs to the user
- Iteration

How does an engineer recognize the need for a product or service?

One of the most helpful skills that engineers have is their ability to make observations. Engineers observe when products, services, or systems are not functioning as they were designed to function. They also create ways to gather data and monitor their innovations to find ways to improve them even when they are working well. Sometimes, engineers are responsible for recognizing needs; at other times, the company may identify the need for a product or service, or a client may bring a need to an engineering team to solve. An engineer's ability to identify the need can determine how successful the project will be. Even if the engineering team is given the need from the client, they are responsible for asking the right questions in order to understand what the user really wants and needs and provide solutions.

What makes a good problem solver?

Good problem solvers are those engineers who are also good at fully understanding the problem. Defining the problem is the most important step of the engineering design process. Many expert engineers spend most of their time on this step! The best way to understand is by asking questions similar to these:

- How much are they willing to spend on this project?
- How does the current solution make them feel?
- What do they want to accomplish when this project is done?

Engineers often face novel technical problems that need innovative solutions. This is why the design process is so critical in tackling a project.

- What do they want?
- Where do they want it?
- When do they want it?

How does an engineer listen to a user and understand their needs?

An engineer has the responsibility to listen to a user and interpret what the user is sharing. They must discover what the unmet user needs are. For example, an online streaming service wanted to know how to get more people to purchase a monthly subscription, so they developed a hypothesis about what would increase membership and then tested the hypothesis with the users by creating test websites. Some other groups use research to develop personas to understand user goals, pain points, and behavior patterns. An equation an engineer might rely on in order to remember how to find unmet needs is: unmet needs = goals + pain points. The goal is what the user wants to accomplish, and the pain point is what is stopping them from accomplishing the goal.

What is the difference between criteria, constraints, specifications, requirements, and objectives?

The design criteria are goals that the solutions must meet in order to be successful. The goals take the form of design specifications from the

DID YOU KNOW?

How does an engineer define a problem?

When people need a definition of a word, they go to a dictionary. When an engineer needs to define a problem, they go to the user to understand their needs and wants. After they have asked many questions, the engineer will translate the user's needs into engineering terms. This is called the problem definition. In order to help the idea-generation process, the problem definition step should not include assumptions that would lead an engineer toward a specific solution. The problem statement must be defined in terms of needs, not solutions.

customer/user. These specifications are unambiguous details that define how the design outcomes are met and how to measure success. Specifications communicate to the user how they should expect the solution (product, system, process, artifact) to perform. From the engineer's perspective, these specifications are viewed as requirements that must be met, and they represent the voice of the user during the design process. Sometimes, the specifications are thought of as user wishes and demands. Constraints are fundamental to engineering design. They are rules and limitations that are put in place and can be used to stimulate creativity and increase innovation. Essentially, constraints are restrictions that force early decisions because options are removed. Sometimes, constraints come from the user, the company, or the engineer's own limited access to funds or resources. If the constraint is not met, the solution is unacceptable to the user. Whenever possible, it is important that specifications and constraints are measurable and expressed in terms of numbers. Although not all constraints and specifications can be quantified, it is helpful for measuring how well a design solution meets the user's needs.

What are some categories of design specifications?

The list of design specifications addresses the aspects of the solutions that the user demands and desires and the constraints that the engineering team must work within. Consider an example design problem:

Design and build a functional model of a vehicle that will travel on Mars terrain and pick up three soil samples of different types.

The user also gave some specific needs (demands): must cost less than $100, must be remotely triggered, the entire device must form a single unit, must

	Sample Specification and Categories	Constraints	User Specification (Demand)	User Specification (Want)
Function (How It Works)	• Performance • Energy (e.g., must operate on inclines up to 45 degrees)		X	
Aesthetics (How It Looks)	• Geometry (e.g., entire device must form a single unit)		X	
Quality (How Well It's Made)	• Materials • Cost • Manufacture • Standards (e.g., off-the-shelf parts and materials should be readily available)			X
Safety	• Time • Transport • Ergonomics (e.g., must be completed during a three-month academic quarter)	X		

be portable, and must pass a safety review. The engineer will take inventory of the specifications (demands and wants) and constraints. Some engineers might create a table to help them organize what they are learning about the problem.

How do engineers do research?

Early in the engineering design process, it is important for engineers to fully understand the problem. Engineers perform research in order to gather the information needed to solve the problem at hand. First, they have to document what they know about the problem; then, they look for missing information that would be helpful to know. At times, it is important for engineers to understand the technology that already exists and to develop improvements. During their research, they will find facts and make inferences based on data, potential users' opinions, and speculations about the problem. For any problem that an engineering team is responsible to solve, they must collect as much information as possible. Gathering information can be done in many ways:

- Ask client experts
- Library research
- Internet research
- Ask other experts

After the engineer has spent some significant time gathering information and doing research, they are ready to begin generating ideas.

How do engineers use brainstorming?

Brainstorming is a technique that requires the engineer to use their imagination to generate a large number of ideas that might lead to possible solutions. This activity is typically completed after the engineers and their team have gathered enough information to get a good understanding of the problem and the needs. They initially develop a wide range of ideas, then identify the most successful solutions to the design problem. This is the solution that will be tested in the next stage of the design process. Brainstorming as a team can be done in many different ways.

What are some strategies to having a productive brainstorming session?

Strategy	Description
Focus on quantity	Generate many ideas and record every idea
Wait to judge	Do not critique any idea during brainstorming
Think outside the box	Wild ideas lead to innovative solutions and designs
Combine and improve ideas	Look across the ideas and determine if you can build upon existing ideas from the team to make new ideas
Keep focus on the problem	Focus the ideas on the need
Set a time limit	Set an appropriate timeline based on the complexity of the problem

What is the ideation process, and how is it different from brainstorming?

The goal of the ideation process is to "go wide" with as many different ideas as possible for a given problem. In addition to brainstorming, ideation has many approaches, including mind mapping, sketching and doodling, creating analogies, brain dumping, storyboarding, prototyping, examining existing solutions, encouraging curiosity by asking questions, making combinations of existing ideas, and so many others.

What are the benefits of the ideation process?

In addition to generating many new ideas, the ideation process is helpful in encouraging engineers to be curious. Curiosity helps engineers ask the right questions and think innovatively. This process can also help them get the most obvious solutions out of their heads and push them to think beyond those solutions. During this process, the engineering design team can learn more about each of the team members' strengths, weaknesses, experiences, and unique perspectives.

How can engineers develop lots of different solutions and concepts?

Heuristics are different strategies that people can use to generate many diverse ideas that are different from each other. They help engineers push the limits of their first ideas and concepts. They were first introduced by mathematician George Polya in his book *How To Solve It*. Some examples of heuristics include:

- Write it down
- Restate it in simpler terms
- Use human-generated power
- Apply existing mechanism in a new way
- Draw a picture

How do engineers communicate their early ideas?

Engineers use prototypes to demonstrate how their selected solutions will function. Prototypes give the target user a fully operational version of the engineering solution. The user can then provide the engineer with feedback on how

DID YOU KNOW?

Why is it important to generate lots of solutions?

Very rarely does the first solution provide the best solution to the problem. In fact, in engineering design, more than one functional solution will always be present. It may be tempting to pick a favorite idea based on your own experiences and data, but it is better to generate as many ideas as possible. To help generate more ideas, people usually work in teams because sometimes old ideas, or other people's ideas, inspire new ones.

well the solution meets their needs and where opportunities for improvement exist. Typically, a prototype is not made of the same materials as the final version. For the prototype, the engineer will use materials that are cheaper and easier to access and manipulate. In addition to the user using the prototype and providing feedback, the engineer can also test the prototype to determine how well it will work and when/how it will fail.

See below for more about prototypes.

How is mind mapping used for brainstorming?

Mind mapping helps both engineers and non-engineers to move ideas and information out of their brains and onto another medium, like paper. It can be an effective strategy to use when experiencing a mental block or idea drought. It is both creative and logical and helps to make connections between the problem being solved and connections to other ideas related to that problem using lines, circles, and links. Connected ideas are shown with lines and create a visual aid that looks much like a tree branch. A mind map can be created in four basic steps:

- Write the key concept as a word or image in the center
- Think about additional important concepts that are connected to the key concept; each one will radiate as a branch from the central key concept and form the main branches
- Think of other concepts that are important to the main branches; each of the main branches can have sub-branches with concepts, which are of lesser importance than the key concept and main branches
- Review your work and see whether any other branches come to mind to add

DID YOU KNOW?

How do engineers present the details of their solutions before prototyping?

Engineers can present their preliminary ideas using sketching and by developing a design specification document. Both of these approaches share details of the design that represent what they have learned about the user's needs and how the solutions will address those needs. The concept sketch is annotated with descriptions of the main functionality of the concept: how it works. Sketching is also a strategy used to generate more ideas. It is okay for a concept sketch to not include dimensions in the early stages, as the goal is simply to communicate the basic elements of the design. This annotated concept sketch can be hand drawn or drawn using computer software. The engineering will label main design elements and include arrows to indicate where these elements can be found on the design. The engineer may also choose to show different views and one- or two-dimensional drawings of the solution concept; this provides as much information as possible to make a decision on the quality of an idea.

Once the engineering team has selected a design solution, they will develop a design specification document that clearly outlines every detail of the problem, the user, the impact of putting the solution in place, major decisions of the solution, and the effort it will take to implement the solution. Because engineers work in teams, this document is very helpful to ensure that the team member who will work on the next phase of the project has all the information they need to be successful.

In the case of using mind mapping for brainstorming, the central concept will represent the problem. The main theme that will radiate as branches will be ideas that the engineer develops and may include existing solutions. The less important or secondary branches that radiate from the new and existing solutions represent interesting ideas, thoughts, or questions that the engineer might want to revisit at a later time.

What is reverse engineering?

Reverse engineering is a technique used by engineers to learn how an existing product functions, about a product's structure and design, and to develop detailed descriptions of products. This technique got its name because the engineer is moving backward through the design process, meaning that instead of starting with a need or problem and then working toward a solution, the engineering team starts with the product and works to understand details about it. Reverse engineering is often used during the information-gathering phase of the process when engineers are trying to understand existing solutions. Another common application of reverse engineering is when a really old part wears out and must be replaced. Due to the age of the part, it may no longer be made, or the company has gone out of business. The engineer will understand how the old part functions and could use a 3D printer to replicate the part that they need. Regardless of the motivation for reverse engineering, the engineer is primarily interested in obtaining the measurements of the physical product and all its elements so that it can be reconstructed using a 3D model. The tools used for reverse engineering will be selected based on the product. Some common tools include:

- Handheld 3D scanner
- Computer-aided design software
- Screwdriver
- Electric drill

How does an engineer choose the best idea if all the top ideas meet the design specifications?

Engineers will use the principles of design to narrow down the generated ideas to one best idea. Some of the principles for evaluating and improving concepts include (1) selecting the concept that can most easily and quickly be built; (2) selecting a design that could perform well when unexpected eternal variations/changes occur; or (3) selecting a design that allows for the main functions to be easily adjusted. Each function of a design should have separate mechanisms for each function and be designed for ease of manufacture.

How does an engineer know that a design concept is easy to manufacture, and what are the benefits of an easy-to-manufacture solution?

An engineer who develops a part that is easy to manufacture can use the time saved to test their design and improve it. Also, a part that is easier to manufacture can also be easier to troubleshoot when issues arise. All of these benefits can also save time and get the solution to the user more quickly. Here are some questions that an engineer might ask when they are deciding if a design idea is easy to manufacture:

Sometimes, several designs might present themselves as possibilities, so a good engineer needs to select the concept that is the most practical and likely to perform well.

- Does it have a large number of parts?
- Does it have many different kinds of parts?
- Do all of the parts need to be made?
- Does the engineer have the skills to make the parts?
- Are the most important functions operating independently of each other?
- Can the design be simplified?

What tools can an engineer use to select the best idea from the alternatives?

A decision-screening matrix allows engineering teams to evaluate the strengths and weaknesses of multiple design alternatives. After a team completes the table, a numerical answer will identify which design alternative is the best when considering the specific design criteria. Sometimes, after a team has completed and identified their best concept, they combine the best elements of several alternatives to create a new solution. Sometimes, the decision matrix can also be used to identify weaknesses in a design. See the example decision-screening matrix below:

Performance Criterion	Weight	Design Concept 1	Design Concept 2	Design Concept 3
Ease of Manufacture	2	3 (X 2)	1 (X 2)	5 (X 2)

DID YOU KNOW?

What are the benefits to selecting a robust design concept?

When an engineer has developed a robust design concept, that means that they know how the concept will respond when small manufacturing errors occur (i.e., the dimensions are slightly off), when the surrounding environment changes within an expected range, and when the design takes on anticipated damage. The engineering team has to consider many different scenarios to improve their design concept to work well when expected changes occur.

Performance Criterion	Weight	Design Concept 1	Design Concept 2	Design Concept 3
Engineering Skill Needed To Build	4	3 (X 4)	3 (X 4)	3 (X 4)
Robust Design	4	5 (X 4)	5 (X 4)	3 (X 4)
Primary Function Independence	3	3 (X 3)	1 (X 3)	1 (X 3)
Total		47	37	37

Scoring: 5 = high, 3 = medium, 1 = low

How does the engineer complete the decision-screening matrix?

Step 1: Identify the alternatives that will be included in the matrix and write those alternatives in the top row of the matrix. They can also be names: design concept 1, 2, 3, etc.

Step 2: Gather the project performance criteria and include them in the first column of the matrix.

Step 3: Consider if any of the criteria are more important than others. If so, assign an appropriate weight to that criterion. The example

above uses a range from 1–5, with 5 representing the most importance and highest-weighted criterion.

Step 4: Develop a scoring system for each criterion; the higher the score, the more positive the result.

Step 5: Rate each of the alternatives according to their score. Multiply each design concept rating by the weight of each criterion.

Step 6: Calculate the total by adding all of the multiplied ratings to the design concept column. Then, select the best concept.

What is a prototype?

After a concept has been selected, engineers need to ensure that the design will work and that it meets the users' needs. Prototypes are functional, scaled-down models of actual concepts that can be tested before the actual design is built. After the prototype is built, it is then tested through experiments and user studies. The engineering team may introduce a prototyped solution into the actual scenario with the user and then collect data on how well the prototyped solution performed. Building a prototype helps the engineering team resolve any questions or uncertainties they may have about the selected design.

What is rapid prototyping?

Rapid prototyping is a technique used to quickly fabricate parts and design concepts through the use of 3D computer-aided design software and a 3D

Prototypes are helpful tools that engineers use to demonstrate, test, or help visualize their ideas for their customers or other engineers.

printer. The use of a 3D printer is called additive manufacturing and is only one type of manufacturing technology. Other manufacturing technologies that are used in rapid prototyping include subtractive manufacturing and compressive manufacturing. With additive manufacturing, material is added layer by layer until the 3D design has been built, whereas with subtractive manufacturing, material is removed by cutting, drilling, or some other technique until the desired shape has been formed. With compressive prototyping, a liquid is forced into a specific shape and then heated until it molds into a solid.

What is a low-fidelity prototype?

A low-fidelity prototype is a quick and simple way to translate a design concept into something that the user can test. Fidelity refers to how realistic the prototype is. Some key elements of a low-fidelity prototype are visual design, inclusion of key design elements, and usability. Some popular techniques include pencil and paper for prototyping digital products, folded-paper and cardboard prototypes, and wood structures and Legos for physical products. A low-fidelity prototype might also include the circuitry components needed in the design.

What is a high-fidelity prototype?

As the selected design gets close to the final version of the product, the fidelity of the prototype increases. A high-fidelity prototype more closely resembles the final product, and the design team can then focus on some additional details of the design. As an example, an engineering team is designing a personal electronic device that will use an electric circuit housed in an enclosure. The low-fidelity prototype may simply include the breadboard circuit with push buttons to test the basic function of the circuit. However, the high-fidelity prototype will solder the circuit parts together and include it within the enclosure that the user will interact with. The entire prototype must be prototyped in different ways at different points of the design process.

How can an engineer test a design?

The design process allows the engineer to find problems and make changes multiple times before they settle on a final design. Once the team has selected a final design, they have to complete one of the most important phases of the process: testing and redesigning. In this step, the engineering team will allow the user to test the final design and provide feedback to the engineering team. Based on the feedback and how the users interact with the design, the engineering team will work to improve the design. Ideally, the prototype will be tested by a user in an environment that is most similar to where the product will be used. For example, if the engineering team is designing a pair of sunglasses, then the testing environment should be a place that is sunny. During the user test, the engineer has the primary responsibility of observing the interaction between the user and the protoype and recording their observations in a notebook. The engineer can also ask questions during the testing stage.

What points are considered by the engineer for the end user?

1. Are your users able to overcome the problem by using or interacting with your solution?
   - If yes, why are they successful?
   - If no, what problems do they encounter that prevent them from being successful?
2. Do the users ever need to ask you any questions when using or interacting with your solution?
   - If yes, what questions do they ask? During what part of their interaction do they ask these questions?
3. Do the users interact with your solution exactly the way that you intended for them to?
   - If no, what do they do differently?
4. If you have measurable targets for your solution, did you meet them?
   - Yes
   - No

How does the engineer know what to redesign once the testing phase is done?

After testing, the engineering team will use all the new information they have gathered to redesign the solutions. This will lead to an improved design concept in which the team fixed issues that came up during testing and refined other aspects of the design. If the prototype was successful in solving the user problem, then the engineer will focus on which aspects made the solution most successful and decide if those should be emphasized more or could be achieved more easily. If the prototype did not solve the user's problem, then the engineering team will address how to remove the aspects of the design that were not helpful. The team will also reflect on the types of questions the user asked and make changes to eliminate the need for those questions. Sometimes, during the testing phase, the user will interact with the prototype in an unexpected way. If this happens, the engineering team will decide if those unexpected actions need to be addressed or corrected. Finally, the solutions can be redesigned by addressing the original criteria and then tested again until the solution is most successful.

What are some strategies for fabricating and manufacturing prototypes?

It is important to minimize the time it takes to get a prototype ready for testing. This is because very few designs work at the first iteration, and new products require many prototypes before they are ready for mass production. Throughout the design process, each design iteration must be tested and modified. In manufacturing, this also means that the machines used to fabricate the design will likely also be modified many times in order to adjust to the changes in design throughout the process. This phase of the process can take much time, and the engineering team must find ways to save time. They can use the following strategies:

- Talk to a machinist, who can give practical advice about the limitations of the machines needed to create the product design
- Do not delay the manufacturing process
- Identify and isolate the separate subfunctions (also known as subassemblies) and delegate those design elements to different members of the team
- Keep detailed drawings current as the design goes through its iterations
- Create and enforce team deadlines
- Always leave adequate time for testing
- Simulate the user conditions as closely as possible

What are some useful hand tools to have accessible for manufacturing?

Some smaller-scale products can be made with common tools.

Hand Tool	Purpose/ Use	Alternatives and Additional Notes
Tape Measure	Measuring length	Alternative: ruler
Calipers	Measuring between the two sides of something	Can also measure inside, outside, and distances between steps

DID YOU KNOW?

How long does product development take?

If the engineering team includes someone with experience in product development and if the product is of moderate complexity, the first prototype can be fabricated within three to six months. A final prototype that both works and looks like a functional product can be developed in six to nine months. The more complex the design or the less experienced the design team is, the more time the product development will take. The fastest time frame that a product can get from concept to the user's hands is one year. Many engineering design teams plan for two years.

Hand Tool	Purpose/ Use	Alternatives and Additional Notes
Level	Determining if a surface is level and balanced	Small, glass tube in the center containing a liquid and an air bubble that indicates a level surface
Protractor	Measuring angles	Multipurpose tool used in engineering and nonengineering settings
X-Acto Knife	Precision cutting	Alternative: utility knife
Coping Saw	Cutting curves and shapes in wood	Blade is thin, narrow, and highly maneuverable
Hacksaw	Making straight cuts in metal	Bow saw makes straight cuts in wood
Scissors	Multipurpose tool used for cutting	Strong and sharp blades are useful for precise cuts
Cordless Hand Drill	Drilling holes in penetrable materials like wood, plastic, or bamboo	Also called a manual drill
Hot Glue Gun	Applying hot glue and joining parts	Need to also acquire hot glue gun sticks before using gun

Hand Tool	Purpose/ Use	Alternatives and Additional Notes
Screwdrivers	Inserting and removing screws	Bits come in many shapes and sizes
Adjustable Wrench	Holding nuts and tightening screws	Cannot grip like pliers; users can adjust the distance between the wrench jaws
Pliers	Grabbing, pulling, bending, and cutting	Different types can work in ways that cannot be done with bare hands
Hammers	Striking surfaces with huge force or removing nails	Can be used to crack or break surfaces; alternative: mallets
Soldering Iron	Connecting wires in a circuit with solder	Gets really hot; apply hot solder to tip of iron (tinning) before heating wires
Wire Cutters	Cutting wires	Typically a built-in function of wire strippers
Wire Strippers	Removing insulation from electric wires in order to make contact	Various notch sizes based on the diameter of the wires; can also cut and shear
Wire Crimpers	Crimping connectors to the end of a cable by deforming pieces and holding them together	Look like pliers; wires should be stripped before crimping

ENGINEERING DESIGN PROCESS AND INNOVATION

What is engineering design?

In engineering, design is an important aspect of problem solving. It is both an action and a way of thinking. The definition of design is not agreed upon by many groups of people. It is often associated with drawing, constructing, creativity, and innovation. Design can also be considered a process or set of activities that bring a concept to reality. Engineering design is different from simply being creative; it involves applying creative ideas, using information to make logical decisions or generate more ideas and creating products or solutions. Engineering design can be applied to problems that are not well defined and problems that do not have easily identifiable starting points.

What is the engineering design process?

Design in engineering involves a series of steps that form a process to solve a problem. The steps, though they appear to be a linear sequence, are often revisited throughout a design experience. Broadly speaking, engineering design processes generally include the following steps:

1. Defining the problem
2. Identifying possible solutions
3. Selecting a solution
4. Implementing the solution
5. Verifying the solution
6. Iterating the entire process

The design process can further be categorized into three design stages and specific activities that one might complete and revisit at each stage. These are problem scoping, developing alternative solutions, and problem realization.

What happens in the developing alternative solutions phase?

In the next stage, developing alternative solutions, the engineer uses the gathered information and begins to brainstorm, test, and assess as many ideas

What is problem scoping?

DID YOU KNOW?

Problem scoping is the design stage in which engineers seek to deeply understand the problem at hand. Experienced engineers spend a significant amount of time at this stage. Engineers who have less experience spend less time at this stage. Some specific activities that engineers engage in during problem scoping are identifying the need, defining the problem, and gathering information. These activities help to ensure that the engineer has the information needed to develop a high-quality solution that meets the needs of the user.

as possible. At this stage, some engineers practice "thinking outside the box" for atypical solutions. The specific design activities in this phase include generating ideas, modeling, feasibility analysis, and evaluation. These activities help the engineering team develop lots of ideas and then use information and testing to decide on a best solution or design concept.

What is project realization?

Finally, in the project realization stage, the engineering team takes a design idea concept and makes it a reality. Often at this stage, the team finds that the final product may not meet the design requirements or could be improved. This is an example of when a team might go back to another design stage and revisit design activities. The critical activities in this stage include design decisions, communication, and implementation. These activities help bring a design concept to a real, tangible product.

What is the difference between criteria and constraints?

When designing a problem solution, engineers need to always be aware of the criteria and constraints of the proposed solution. Criteria are requirements that have a sliding scale, and a range of acceptable options exist. In most cases, some of the presented options are better than others. Broadly speaking, criteria are principles the engineer follows as they build or design something to make sure it is going to work the way they want it to. In other words, what will it look like and how will it work? The engineer has to also think about things like shape, size, weight, speed, and how easy it might be to make.

Some questions they may think about are:

- Does the solution need to be aesthetically pleasing?
- Should the solution be close to a target cost?
- How easy is the solution to implement?

On the other hand, constraints are requirements that are either met or not met. The answers to these are yes or no. For example, constraints are limits to what the engineer can do as they build or design something. Some common ones are cost, time, place, and how much the engineer knows.

The engineer might seek to answer these questions:

- Can the solution be implemented in a certain amount of time?
- Does the solution have a minimum or maximum of some measurable quality?

What contextual features can influence engineering solutions?

In engineering, the role of context on the design and implementation of engineering solutions cannot be understated. In fact, when engineers define problems, the first step in the problem-solving process, they must take into account how the context in which the problem exists may influence the solution. Several contextual features may impact the solution that the engineer designs. These are:

- Global features: The engineer must consider what geographic and cultural conditions may influence the design choices in solving the problem.
- Economic features: The engineer must identify the economic conditions that may affect finding a solution. For example, an engineer may consider more or less expensive materials based on the economic status of the target users and stakeholders.
- Environmental features: The engineer should seek to answer the question of what environmental impacts, planned or unintended, result from the material, manufacturing process, and life cycle.
- Societal features: The engineer should clearly identify the societal impacts that were considered during the development of the solution.

Why do engineers create context maps?

Creating a context map can be a useful aid when trying to think about all of the different factors that affect a problem or potential engineering design.

Context maps also help the engineer to identify potential factors that could influence the overall design.

- Disciplinary influences: What types of disciplines (both inside and outside of engineering) have knowledge about the problem and should be included in the problem-solving process?
- Organizational influences: What organizational structures exist that would influence the project's potential solutions? These could be governmental organizations, academic organizations, professional societies, or overall societal structures.
- Regulations: What types of laws and regulations exist that need to be taken into account when solving the problem?
- Cultural norms: What types of societal and cultural beliefs should be taken into consideration? What ethical considerations need to be included?
- Specific expectations: Do stakeholders have any specific expectations that would influence the type of solution the engineer tries to find?

Who is a stakeholder?

A stakeholder is someone who has some vested interest in a project. This person or entity is and can be affected by the system in any particular way. The stakeholder is also capable of affecting the system as well. For example, if designing a playground, some of the stakeholders would be parents of the children using the playground, the children using the playground, local and state government entities and officials, investors, or designers of the playground equipment.

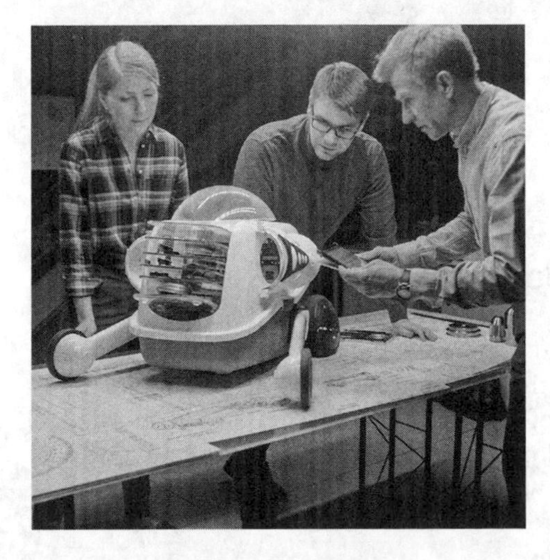

Innovations can come from any field of engineering, not just the high-tech industry. Robots and AI are cool, but important progress can be made by innovating systems and processes, as well as objects that might not be high-tech at all.

DID YOU KNOW?

How does engineering create knowledge?

Another aspect of innovation is creating knowledge as opposed to only learning about engineering in school. The future of engineering requires that engineers not just simply learn about engineering and science concepts but also build upon that knowledge by developing solutions that are cutting edge and meet current and future needs of the society.

What is innovation?

Innovation can be an object that was created and impacts society or meets a need in an unexpected or new way. It can also be the application of a process or product in order to benefit a domain or field. This means that it is a way of thinking that allows someone to create solutions and products that benefit others. The word innovation may trigger an immediate association with creativity or with being creative. However, innovation carries an additional meaning with respect to how an engineer uses creativity and innovates a process, system, or object that benefits society or a group of people.

What are the four characteristics of innovation?

- Collaboration
- Learning across boundaries
- Trial and error
- Creating knowledge

What is important about innovation?

Innovation values collaboration over individual achievement. Often in traditional problem-solving scenarios, the problem solver might work alone to develop a viable solution. The culture of innovation reinforces that the knowledge, skills, and processes needed to develop the best solution do not exist in one person but will require many people to work together or collaborate to develop the best solution.

What does innovation require?

Innovation requires knowing how to work and combine knowledge from different disciplines to understand and solve problems. When an engineer applies an innovative process, they use all available knowledge and resources, including knowledge from other disciplines (i.e., engineering, science, business, history). Through collaboration, teamwork, and classes in different nonengineering fields, engineers develop knowledge related to different areas to solve the complex problems that our society faces.

What is the relationship between trial and error and innovation?

In innovation, the practice of trial and error is valued over that of risk avoidance. A culture of trial and error encourages engineers to fail early and fail often. Risk avoidance can reduce the impact of design solutions. Often, failing early and often leads to many improvements and iterations in the design process until the engineer and their team reach the best solutions.

What is the difference between innovation and engineering?

Engineers can be innovative and can also produce innovations. Engineering is focused on identifying problems and developing solutions or developing solutions to given problems. Engineers create innovations that impact society. Engineers also innovate during the design process to create products, systems, and objects that impact society.

What are some of the top innovations of the twentieth century?

History is really important! In engineering, it is important to understand the engineering accomplishments of the past so that we can advance the work and impact of these technologies. Twenty engineering achievements of the twentieth century are listed below:

- Electrification
- Automobiles

Optical fibers (or fiber optics) have been a boon to communication. Fiber optic cables experience less signal loss than metal (e.g., copper) cables and they are not vulnerable to electromagnetic interference.

- Airplanes
- Water supply and distribution
- Electronics
- Radio and television
- Agricultural mechanization
- Computers
- Telephones
- Air conditioning/refrigeration
- Interstate highways
- Space flights
- Internet
- Imaging
- Household appliances
- Health technologies
- Petrochemical technology
- Laser and fiber optics
- Nuclear technologies
- High-performance materials

For more information on each of these innovations, engineers who worked on innovation, and its timeline, read *The Century of Innovation*, a book by the National Academy of Engineering.

DID YOU KNOW?

Where do engineers solve problems?

Engineers are trained to find problems and solve problems that are given to them. They do this work in many settings (in the office, at work sites, on the computer, with teams, etc.). Most often, the problems engineers solve are specific to a particular context, group of people, region, etc.

What kinds of problems do engineers solve?

Engineers are good at solving complex problems that are not easily defined and have many different solutions (ill-structured problems). They are also well trained to solve problems that can be solved by applying the correct algorithm or process (well-structured problems). Ill-structured problems are sometimes described as messy and often do not have a clear starting point. They most resemble real-world challenges and issues—for example, problems like plastic pollution, global warming, educational inequalities, or trash disposal. These are big problems with aspects that include processes, people, and policies. Solutions to these types of complex problems require teams with many different disciplinary backgrounds. Engineers will work on complex problems together with non-engineers to develop and implement solutions to complex problems.

Do engineers solve well-structured problems?

Engineers can also solve well-structured problems that may not require team members or those with diverse skill sets. Well-structured problems can be simple or complex. They most resemble textbook problems. These types of problems are still prevalent in engineering work, and they allow for engineers to apply their specialized knowledge to provide improvements to existing solutions. For example, in a manufacturing plant, a part may be found to be repeatedly defective. This can be a challenging problem because it involves the part itself, the manufacturing process, and the individuals who are on the line building the part.

What types of ethical standards must engineers adhere to?

Most engineering disciplines are governed by their specific governing bodies and, as such, have discipline-specific ethical standards. However, the National Society of Professional Engineers, NSPE for short, has identified six fundamental canons, i.e., principles, that ALL engineers are expected to adhere to. These are:

1. Engineers should always prioritize the safety, health, and welfare of the public.
2. Engineers should only operate in areas where they are highly competent.
3. Engineers should be truthful and objective in all their communication with the public.
4. Engineers should work exclusively for each employer or client for the entire duration of their project.
5. Engineers should not be deceptive about their qualifications, knowledge, work experience, or abilities.
6. Engineers should conduct themselves honorably, responsibly, ethically, and lawfully so as to enhance the honor, reputation, and usefulness of the profession.

This code of ethics should be adhered to at all times regardless of the perceived consequences. They are also used to guide engineering rules of practice and professional obligations. This means that embedded in the code of ethics are specific rules that engineers must follow. No matter what the problem is or in what context the problem is found, the engineer is expected to act in an ethical manner.

What is an engineering way of thinking?

An engineering way of thinking is the practice of using one's knowledge of mathematics, science, and technology as tools when solving multistructured problems. Through their education and training, engineers develop ways of viewing problems, skills in applying their technical knowledge, and critical thought to devise solutions that are appropriate for the specific context in which the problem exists.

How important is measurement and accuracy in engineering?

Measurement and accuracy are two very important concepts in engineering. When engineers design and create any product, the quality of the product is often determined by its ability to meet the customer's needs. If a part is not accurately measured, it can derail the entire process and the finished product as well.

Is communication important for engineers?

Yes. It is extremely important that engineers are able to communicate effectively not just with other engineers but with the public in general. In their jobs, engineers must work with all personnel, some of whom may be non-engineers such as production supervisors, designers, creative writers, marketing managers, etc. An engineer must be able to communicate his design ideas and solutions in such a way that everyone is able to understand and make meaning of the information.

What grand challenges must engineers be prepared to face?

In 2008, the National Academy of Engineering (NAE) invited a group of panelists who were deemed leaders in technological thinking and innovation to

The National Academy of Engineering is part of the National Academy of Sciences based in Washington, D.C.

create a list of goals that are needed to improve the lives of everything and everyone on our planet. These thought leaders came from all over the world and worked to identify what was called the Grand Challenges for Engineering in the 21st Century. The list consists of 14 challenges that were then categorized into four main categories: Sustainability, Health, Security, and Joy of Living.

Sustainability

- Make solar energy economical
- Provide energy from fusion
- Develop methods for carbon sequestration
- Manage the nitrogen cycle
- Provide access to clean water

Health

- Advance health informatics
- Engineer better medicines

Security

- Prevent nuclear terror
- Secure cyberspace
- Restore and improve urban infrastructure

Joy of Living

- Reverse-engineer the brain
- Enhance virtual reality
- Advance personalize learning
- Engineer the tools of scientific discovery

What special skills are engineers expected to have?

For most jobs, engineers are required to have certain skills. Some of the most common skills engineers are expected to have are:

- Creativity: This is where the engineer is able to think outside of the box. As they complete work-related tasks, the engineer must be able to ask the right questions, consider new ways to solve problems, or create new systems that can be applied to the given context.
- Problem solving: Being able to solve complex and ill-structured problems are skills at the core of being an engineer. To this end, the engineer

One challenge engineers are working on for the future is improving virtual reality technology.

must be able to identify and define the problem, conceptualize possible solutions, test those ideas, and converge on the most appropriate and most applicable solution.

- Teamwork: Engineers almost never work alone. Most times, they work with other engineers of the same or different disciplines or non-engineers such as marketing managers, designers, technicians, etc. It is important that the engineer knows how to work collaboratively and effectively with others in order to meet their intended goals.
- Disciplinary knowledge: Engineers must be able to understand the fundamental and complex concepts of their discipline. This understanding is very important when they seek to design, improve, or use existing processes to complete their jobs.
- Strong analytical skills: Like disciplinary knowledge, it is also important that engineers know how to use and apply complex mathematical and scientific understandings, such as calculus, or differential equations, when designing, analyzing, and troubleshooting.
- Communication: Engineers must be able to communicate effectively. When they create problem solutions or conduct analyses, it is important that engineers are able to communicate their findings and recommendations in a clear, concise, and meaningful way.
- Leadership: In most cases, engineers will be called upon to lead. They must possess good leadership skills so that their team can function effectively.
- Project management: Like leadership skills, engineers are also expected to be able to manage multiple projects of varying complexities simulta-

neously. This involves knowing the availability of resources, materials, and personnel necessary to complete the project.

- High ethical standards: Engineers must always adhere to and know how to apply the professional code of ethics to their job and their interactions with the public.
- Professionalism: It is expected that the engineer maintains the highest levels of professionalism at all times.
- Lifelong learning: With continued changes in technology and how they shape the world we live in, it is critical that engineers are able to adapt to changing times. An engineer is expected to constantly learn new technologies, new approaches to problem solving, and new ways of engaging with the global world.

Do engineers ever make mistakes?

Yes. While engineers have designed and created some of the most technologically and scientifically advanced systems in the world to make our lives easier and better, sometimes, things did not go as planned and resulted in serious damage, even loss of life. Here are some of the top engineering failures that resulted in catastrophic disasters across the world:

- Sweden: *Vasa* sinking (1628): This warship sank, killing all 30 people aboard. The failure was attributed to the amount of weight the ship was

The *Vasa* was a military ship commissioned by the King of Sweden. It took two years to build and was the most elaborate and well-armed naval vessel of its day when it launched in 1628. It promptly sank 1,400 yards into its maiden voyage because the construction was so unstable, resulting in great embarassment for the king and a big financial loss, too. (The *Vasa* was recovered in 1961 and now is on display at the Vasa Museum in Stockholm.)

carrying in and on its upper levels, which led to it being very unstable. It sank shortly after setting sail.

- USA: SS *Sultana* fire (1865): Known as the worst marine failure in U.S. history, this side-wheel steamboat had three of its boilers explode, which caused it to burn down. This disaster was particularly deadly because the ship was well over its stated limit of 376 passengers; it was instead carrying well over 2,000 people (2,137, to be exact). The number of passengers who died was estimated to be 1,238.

- USA: Johnstown, Pennsylvania, flood (1889): When severe rain fell upstream of the town, delivering 14.55 million cubic meters of water, over 2,200 people died. This disaster was caused by dam failure due to poor maintenance and extensive rainfall just a few days prior to the flood, which almost wiped the town off the map.

- UK: RMS *Titanic* sinking (1912): Called the "Unsinkable Ship" at the time of its voyage, the *Titanic* ran into an iceberg on its very first trip and eventually sank due to excessive flooding. The exact number of passengers who died is still unknown, but the current estimated total is 1,500.

- Canada: Quebec Bridge collapse (1907, 1916): Known as the longest cantilever bridge in the world, this system failed twice. First, in 1907, the bridge collapsed because the frame was not designed to withstand the full weight of the bridge. The second time the bridge collapsed was in 1916. This time, the cause of failure was attributed to the center span becoming dislodged as the bridge was being fitted in its place. Across the two failures, 88 people were killed.

- UK: Gretna rail crash (1915): When two signalmen failed to realize that the northbound train was running on the southbound lane, both trains had a head-on crash that resulted in the deaths of 226 people. This is still known as the worst railway crash in British history.

- USA: Boston molasses flood (1919): When a storage tank containing 2.3 million gallons of molasses burst due to thermal expansion of the molasses, a river of molasses going 35 miles per hour ran through the streets, killing 21 people and causing injury to 150. Nearby buildings sustained significant damages, too.

- USA: St. Francis Dam failure (1928): Designed as a curved, concrete gravity dam, this reservoir was designed to meet the growing need for water regulation and supplied water to Los Angeles, California. When the dam burst, the water caused the deaths of 431 people. The failure was attributed to poor design and low-quality concrete.

- USA: *Hindenburg* crash (1937): Thirty-five people died when this German airship caught on fire and crashed as it attempted to dock on its station. One person on the ground was also killed as a result of this. The cause of the fire was attributed to an electrostatic discharge that ignited hydrogen gas that was leaking.

- USA: Cleveland gas explosion (1940): When the weld bead of a storage tank failed and started releasing liquefied natural gas, it caused a chain explosion, killing 130 people and damaging homes and businesses within a square mile of the gas plant.
- USA: Tacoma Bridge collapse (1940): This four-month-old bridge collapsed into the water when the main span experienced irreparable damage. Fortunately, no lives were lost. The failure was attributed to aeroelastic flutter. The bridge, having solid sides, was not designed to allow wind to pass through, and this further compounded the problem.
- USA: Apollo 1 fire (1967): A short circuit in the wiring of the aircraft led to the deaths of all three astronauts when a fire broke out in the pre-flight test. This mission was intended to be the first crewed mission of the Apollo space program.
- USA: Apollo 13 explosion (1970): Damaged wire insulation mixed with oxygen caused an explosion of the tank. Because of the explosion, the shuttle orbited the moon but landed safely in the Pacific Ocean. Fortunately, all three astronauts in this shuttle walked away with no injuries.
- USA: Skylab disaster (1974): No one was injured when the first U.S. space station orbit decayed and crashed to Earth. Some of the pieces fell in Australia because of errors in the calculations.
- China: Banqiao Dam failure (1975): This dam, once called the "Iron Dam," was known as an engineering wonder until it burst after Typhoon Nina caused severe rainfall. The collapsed dam led to a 20-foot-high wall of water about 7 miles wide rushing through towns and villages. This disaster remains one of the deadliest in human history, with 26,000 people dying as a direct result of the flood. However, the estimated death toll is believed to be between 171,000 and 240,000 due to ensuing conditions after the flood, such as the lack of food, clean water, and medical attention from being stranded.
- USA: Flight 191 DC-10 crash (1979): This American Airlines flight crashed before fully taking off when one of the engines detached, thereby cutting several main lines on the vessel. This resulted in the deaths of everyone aboard—271 passengers and crew members as well as two ground workers.
- USA: Hyatt Regency Hotel disaster (1981): When two walkways (one directly above the other) collapsed, 114 people were killed and another 216 severely injured. This failure was caused by flaws in the steel rods used to support the walkways.

- India: Bhopal leak (1984): It is estimated that 16,000 people died and well over 500,000 were exposed to a very toxic and deadly gas (methyl isocyanate, or MIC) when a pesticide plant experienced a leak. To this day, the area is still contaminated, with the water and land near and around the site declared unsafe for humans and animals.
- USSR: Chernobyl nuclear explosion (1986): Several people, estimated between 30,000 and 60,000, died of cancer or related complications when a reactor at this nuclear plant initiated a set of events that resulted in an explosion. Radioactive contaminants were released into the air for nine days before the fire could be contained.
- USA: Space Shuttle *Challenger* explosion (1986): This tragedy took place on live television and was the result of an external tank exploding when the rocket booster came loose, causing a rupture in the tank just 73 seconds after the vessel took off. All seven astronauts were killed.
- France: Air France Concorde crash (2000): After 24 years of passenger service, this vessel crashed after blowing out a tire during takeoff. A piece of the tire flew into the fuel tank, which led to the tank rupturing, causing fuel to leak and, subsequently, a fire in the engines. All 109 people on board, as well as four people on the ground, were killed.
- USA: Space Shuttle *Columbia* disaster (2003): When this spacecraft attempted its reentry into Earth's atmosphere, it completely disintegrated due to compromised foam insulation, which likely happened when the shuttle first took off. All seven crew members died.
- France: Charles De Gaulle Airport collapse (2004): A portion of the roof in one of this airport's terminals collapsed, killing four people and

This photo shows Reactors 1 through 4 (left to right) at the Fukushima Nuklear Power Plant shortly after the 2011 earthquake and tsunami that caused a meltdown and radiation leak that will plague Japan for generations to come.

severely injuring three. The failure was attributed to several design flaws plus poor concrete and steel.

- South Korea: *Deepwater Horizon* explosion (2010): Known as the deepest oil well, a drilling explosion resulted in the deaths of 11 people. The entire rig sank a few days after the explosion, causing the largest marine oil spill in history. This failure was attributed to an uncontrolled release of crude oil from the well.

- Japan: Fukushima Daiichi nuclear disaster (2011): When the reactor core failed to cool down rapidly following explosions caused by an earthquake, radioactive contaminants were released into the air. Fortunately, no one died nor were any injuries reported.

How does an engineer iterate on a design during prototyping and testing?

During the design process, engineers continue to gather information as the use interacts with the prototype. The engineers can use the user feedback to determine how closely the prototype met the users' needs and expectations. Testing of the prototype can occur with the user or through experiments. During experiments and trials, the prototype can be tested to determine under which conditions it will fail. This information will tell the engineer if the prototype meets the design criteria developed earlier in the design process.

How do engineers present their work?

Typically, engineers present their work throughout the design process in different forms. Technical communication is something that engineers must develop so that they can provide both engineers and non-engineers with useful and easy-to-understand information about their engineering work. During the design process, engineers present their work as they meet specific milestones. For example, in the earlier phases of the design process, an engineering team will present a summary of their gathering information about the given problem and provide potential solutions during a preliminary design review (PDR). In a PDR, the engineers will orally present what they have learned, share their ideas and sketches, and receive feedback from their team, other teams they are working with, and even the user they are designing for. Once their ideas are reviewed, they will make improvements on their potential solutions, develop prototypes of the selected ideas, and test those prototypes. Once the engineering team has selected the best concept, they will develop and test a prototype and then present their design details and prototype in a critical design review (CDR). In a CDR, the engineering team has the responsibility of convincingly demonstrating to their team, other teams, and the user that they have developed a functional concept that meets the design criteria. In addition to oral presentations, engineers also develop reports and documentations that detail every aspect of their chosen design. Engineers present their work through presentations, reports, prototypes, and detailed drawings.

What are examples of engineering design problems?

The engineering design process can be used to solve problems that are very simple or those that are very complex. When an engineer is learning about the problem they are solving, they try to develop very short statements that describe the problem. Here are some examples of engineering design problems:

- A teacher needs a way to deep clean whiteboards and save time. Here is the problem statement: Teacher A wanted an automatic solution that would be as simple as a light switch. Cost was a constraint. It needed to be cheap. For this teacher, a successful win would look something like a clean whiteboard.
- Farmers in Bangladesh and the United Kingdom need a better way to grow crops. Here is the problem statement: Farmer A has been experiencing challenges growing crops on land that is regularly flooded. He needs a new method to grow crops that is not completely dependent on having a stable soil location.

EDUCATIONAL INSTITUTIONS AND RESEARCH CENTERS FOR DESIGN THINKING AND ENGINEERING DESIGN

- Berkeley-Haas Innovation Lab
- ArtCenter College of Design: Designmatters
- Massachusetts Institute of Technology: D-Lab
- Northwestern University: Segal Design Institute
- University of Texas at Austin: School of Design and Creative Technologies
- Stanford University: d.school
- Delft University of Technology: School of Design

- University of Washington: Human Centered Design & Engineering
- University of Michigan at Dearborn: Human Centered Engineering Design
- University of Michigan: Design Science

FAMOUS PEOPLE IN ENGINEERING DESIGN

Who is Clive Dym?

Clive Dym (1942–2016) was an engineering design professor and researcher and also served as the director of the Center for Design Education at Harvey Mudd College. He earned a BS from Cooper Union in 1962, an MS from Polytechnic Institute of Brooklyn in 1964, and a Ph.D. from Stanford University in 1967. He was a strong advocate for the integration of design, making, and prototyping in engineering classroom instruction. This helped students learn about manufacturing and design and how to communicate about their work. Dym believed that design was the distinguishing ability and activity of an engineer. Outside of the classroom, he also developed a series of workshops that brought together scholars and educators who are committed to design and engineering education. They are now named in his honor: the Clive L. Dym Mudd Design Workshops. He received many honors, including the National Academy of Engineering Gordon Prize for creating and disseminating innovations in undergraduate engineering design education for the development of engineering leaders.

Who developed the concept of design thinking?

Tim Brown is a British industrial designer who has been a trailblazer in the development and introduction of design thinking to many different audiences and industries all over the world. He is considered a top influencer in both the United States and United Kingdom, and he also served as the chair of IDEO, a global design company at the forefront of the field. IDEO, the organization he chaired, became famous for practicing design thinking and applying it to solve as many different types of problems as possible. Design thinking was first introduced in the 1960s when researchers were trying to understand creativity techniques and how to evaluate, design, and solve problems. To date, design thinking does not have one best definition, but Tim Brown offers this definition, which many people use as a starting point:

"Design thinking is a human-centered approach to innovation that draws from the designer's toolkit to integrate the needs of people, the possibilities of technology, and the requirements for business success."

His approach to design thinking is that it can be an idea, a strategy, a method, or just another way to see the world. He worked closely with IDEO founder David Kelley, who helped bring the idea of design thinking into business and innovation sectors.

Who Founded IDEO?

David Kelley (1951–) and Tom Kelley (1955–) are brothers who together are the face of the IDEO design firm. David is a businessman, entrepreneur, designer, engineer, professor/teacher, and founder of IDEO. He is an active professor at Stanford University and has been recognized for his contributions to design and design education. He earned a bachelor's degree in electrical engineering and a master's degree in design. He is also known for creating the world-renowned d.school at Stanford and has been awarded honorary doctorates for his contributions. Tom Kelley is a partner in the IDEO design firm and currently has a global role in Japan. The Kelley brothers are best-selling authors of *The Art of Innovation*, a powerful and compelling book on unleashing the creativity that lies within each and every person.

Who is Linda Hill?

Linda Hill (1957–) is a Harvard professor who investigates innovative cultures within different types of organizations. She is revered as a top scholar

An engineering professor at Stanford University who also founded the Hasso Plattner Institute of Design there, David M. Kelley is cofounder of the IDEO design firm in San Francisco.

DID YOU KNOW?

Who is Tahira Reid Smith?

Tahira Reid Smith (1977–) is an American engineering professor and inventor who is best known for her automated double Dutch jump rope machine. She grew up in the Bronx, New York, and was encouraged since she was young to be inventive, creative, and keep an interest in mathematics by her Jamaican American family and her teachers. Reid received her first patent for the double Dutch device in 1999 (U.S. Patent 5,961,425) and a second (U.S. Patent 6,634,994) in 2003. She earned bachelor's and master's degrees in mechanical engineering from Rensselaer Polytechnic Institute (RPI) and a Ph.D. in design science from the University of Michigan. Her RPI professor, Burt Swersey, taught her Introduction to Engineering Design course and later became a mentor and co-inventor.

in the areas of leadership and innovation. She applies design thinking to leadership and organizational development in the business world. She helps organizations make the best use of diverse thinkers to solve challenging problems. She believes that innovation is the power of an organization's collective genius.

Who is Cynthia Atman?

Cynthia (Cindy; 1958–) Atman is an American scholar and engineering educator. She is the founding director of the Center for Engineering Learning & Teaching (CELT) and a professor in the Department of Human Centered Design & Engineering at the University of Washington. She is a predominant and leading engineering education scholar who was able to translate how to use elements of the design process to teach design to engineering students. She and her collaborator, Jennifer Turns, identified specific design behaviors of designers ranging from novice to expert levels.

Electrical Engineering

BASICS

What is electrical engineering?

Electrical engineering is the engineering division that deals with electricity technologies. It is concerned with the research, design, and implementation of electricity, electronics, and electromagnetism machinery, devices, and systems. A wide variety of components, instruments, and devices are used by electrical engineers from tiny microchips to gigantic power station machines.

What is the difference between electrical engineering and electronic engineering?

Electrical engineering deals with the production and delivery of electricity to a particular location or equipment on a broad scale. Electrical engineers handle the design, development, testing, and supervision of electrical equipment production, particularly the manufacturing of electric motors, radars, and navigation systems. However, electronic engineers develop the internal circuits of cell phones, audiovisual devices, televisions, spacecraft, and radars.

DID YOU KNOW?

What types of problems do electrical engineers solve?

Electrical engineers design and construct new and innovative electrical devices. They are responsible for fixing issues and monitoring equipment. They also apply the principles of electricity, electromagnetism, and electronics to data-processing and energy-transmission systems on large- and small-scale systems. Thus, electrical engineers have the responsibility of generating, transmitting, and distributing electric power.

What types of jobs do electrical engineers do?

- Design new approaches to build or optimize products using electrical power
- Product database management by writing computer programs and entering data
- Use computer-aided design (CAD) systems to create schematics and circuits
- Establish standards for manufacturing, construction, and installation
- Assessment of electrical devices, products, parts, and systems
- Use computers to simulate how electrical devices and systems will function
- Design electrical components by researching consumer demands
- Manage electrical project production to ensure that work is done well, on schedule, and under the budget
- Improve production processes by improving machine design and modification
- Analyze and test finished goods and device functionality

What skills should an electrical engineer have?

- Design of circuits
- Apply theory and analysis of linear systems
- Build electric schematics
- Gather and analyze data

- Create programmable-logic controllers (PLCs) for industrial applications
- Develop and use computer systems
- Maintain electronic equipment
- Prepare operational plans
- Project management
- Design systems for efficient and reliable operations
- Identify and solve engineering design problems
- Use Autodesk and AutoCAD specifically for electrical control systems
- Be proficient in MATLAB
- Have knowledge of raw materials, quality control, techniques, and production processes

What tools do electrical engineers use?

- Frequency calibrators and simulators
- Laboratory evaporators and safety furnaces
- Oscilloscopes
- Scanning probe microscopes
- Voltmeters or multimeters
- Semiconductor process systems (for example: electron beam evaporators, wafer steppers, and wire bonders)
- Signal generators
- Calculators
- Spectrometers
- Tube furnaces
- Voltage or current meters
- Electrodes

ELECTRICITY CONCEPTS

What is electricity?

Electricity is the physical manifestation of or a form of energy that often results from charged particles' existence and interaction. Electricity occurs when charged particles move between positive and negative regions within a conductor.

Today, electricity is widely used to power everything from small personal devices, such as cell phones, to entire cities. In the world around us, electricity is a natural occurrence and can be observed in lighting. A well-known experiment was conducted by Benjamin Franklin, who went flying his kite during a thunderstorm in 1752. Using a kite with some conductive materials (e.g., copper) and nonconductive strings (e.g., silk), Franklin discovered that lighting was electrical discharge.

What is the difference between static electricity and current electricity?

Static electricity develops on the substance's surface. It is generated by friction or instant interaction of negative charges from one object to another such as rubbing two materials against each other. Static electricity occurs on objects isolated by an insulator when a buildup of opposite charges occurs.

Current electricity occurs when charges are allowed to continually flow (movement of electrons) in a conductor. Batteries make up the most widely used source of current electricity.

What are the different methods of producing electricity?

Power Type	Method
Thermal Power	Coal and natural gas are used by combustion

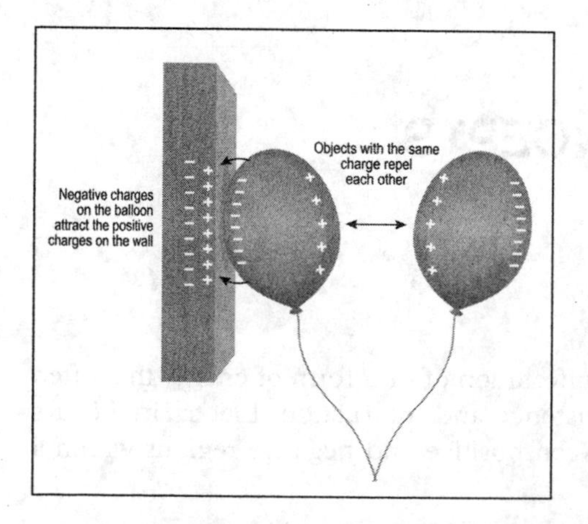

Figure 1: Rubbing a balloon will cause an imbalance of electrons on its surface; the charged balloon will then become attracted to the opposite charge on another surface.

DID YOU KNOW?

What is dielectric strength?

Dielectric strength is the voltage that can be endured by an insulating material before breakdown takes place. Typically, it depends on the material's thickness and the testing method and conditions. It is defined as measuring a material's dielectric breakdown resistance under an applied voltage and is expressed as volts per unit thickness. Thus, it is a measure of how strong a material is as an insulator.

Power Type	Method
Nuclear Power	Uranium is used by nuclear fission
Photovoltaic Power	Sunlight is converted into electricity
Wind Power	Rotation energy by wind is converted into electricity
Hydro Power	Rotating water wheel is used by running water
Geothermal Power	Magmatic heat boils underground water to rotate the steam turbine to generate electricity
Friction Power	Static electricity is produced
Oxidation Power	A chemical reaction in cells and batteries
Mechanical Power	Electricity is produced a by generator

What factors does dielectric strength depend on?

Factors affecting dielectric strength are:

- Type of plastic and electrodes
- Size and shape of the plastic and electrodes
- Rate with which the field is increased
- Medium that surrounds the insulator
- Sample thickness
- Operating temperature
- Form or distribution of the field of electric stress in the material

- Frequency, rate, and duration of the applied voltage
- Fatigue with repeated voltage application
- Moisture content
- Chemical changes under stress

What is an electric charge?

This is the essential physical property of matter demonstrated by electrons and protons and is the fundamental electricity quantity. An electric charge is the force exerted on a particle when placed in an electromagnetic field. The electromagnetic field is a combination of an electric field created by an electric charge and a magnetic field generated when the electric charges move. Electric charges can be either positive or negative. At the atomic level, positive charges are carried by protons, while electrons carry negative charges. The symbol Q denotes the charge. The unit of charge is known as a coulomb, represented by the letter C, and is the amount of charge transferred by 1 ampere of current in one second. This unit is named after renowned French physicist Charles-Augustin de Coulomb.

What is voltage?

Voltage, also known as electromotive force (EMF) and electric potential difference (PD), drives a uniform flow of electrons through a circuit. The voltage in a circuit is provided by a source and is, therefore, the specific measure of potential energy relative between two points. Some sources are fixed, meaning they remain constant, e.g., a battery, while others are variable and, hence, the value of the voltage input can change. In a circuit, the voltage causes the charges to move through the conductor, creating an electric current. The voltage unit is volts, represented by the letter V, and is named after Italian physicist Alessandro Volta, who made the first electric cell in 1800.

What is current?

Current is the movement of free electrons in a closed circuit when charges travel through the material. These free electrons only move when a potential difference, i.e., voltage, is applied to them. Therefore, the magnitude of the current is determined by the strength of the potential applied at each end of the conductor. The electric current unit is ampere, represented by the letter A, and is named after French physicist and mathematician André-Marie Ampere. The ampere is 1 coulomb of charge per second.

DID YOU KNOW?

What is Eddy current?

According to Faraday's law of induction, Eddy currents are electric current loops induced (circulated) inside conductors by a changing magnetic field in the conductor that flow in closed circles perpendicular to the plane of the magnetic field. They can be created when a conductor passes through a magnetic field or when a stationary conductor's magnetic field changes. Thus, it is the current generated in the interior of conducting masses by variation of the magnetic flux.

Two types of current directions exist. The most commonly discussed is electron flow, which is the movement of electrons from negative to positive. The other direction is known as a conventional flow, which is the movement of holes (positively charged particles) from positive to negative. The current flow is significant for circuit calculations because the type of flow denotes the symbol (positive + for conventional flow and negative – for electron flow) that should be added to the current value. The electric current moving through a conductor always produces a magnetic field.

What is direct current?

Direct current, or DC, is a type of current that changes in magnitude, meaning that the current's overall intensity can vary with time. However, the direction of the current flow never changes. DC is also referred to as zero-frequency or constant current. DC is obtained from sources that provide a constant current such as voltaic batteries, solar or fuel cells, or DC generators. DC can also be generated from alternating-current sources by the use of semiconductor devices known as rectifiers. DC is most commonly used for digital electronic devices as well as other small, electronic devices.

What is alternating current?

An alternating current, or AC, is a current type that changes in magnitude and direction periodically. This means that the current value goes from zero to its maximum in one direction, decreases to zero, increases to its maximum in the opposite direction, then goes back to zero. As the name suggests, the value of the current alternates between zero and some given maximum and direction,

Entrepreneur George Westinghouse fought a long battle in the 1870s and '80s against Edison with his Westinghouse Company championing AC current. After a couple accidents involving electrocutions from AC power lines, Edison's DC system won the day.

i.e., positive or negative. The alternating current was developed by Nikola Tesla, a Serbian scientist. Most commonly, AC is used to provide electrical current to homes and other buildings because of its ability to be transmitted over long distances at high levels. Unlike DC, the voltage value can be changed, increased, or decreased using a power transformer.

How do you convert DC to AC?

Sometimes, it is necessary to convert alternating current to direct current. This is done through a process known as rectification. Rectification is achieved using electronic devices that permit the flow of current in one direction only. These devices are made from semiconductor materials that will act as a conductor for the positive half of the current cycle while working as an insulator for the cycle's negative half. The device that converts AC to DC is known as a rectifier.

What is the power factor?

An AC electrical power system's power factor is defined as the ratio of the actual working power consumed by the load to the apparent power (demand) flowing in the circuit. Actual working power is measured in kilowatts

What is the difference between 120V and 240V electricity?

A standard electrical outlet contains a 120-volt wire and a neutral wire, which deliver power using one electrical service phase. For example, microwaves, refrigerators, and dishwashers require a 120-volt supply, whereas 240-volt outlets use two 120-volt wires simultaneously, plus a neutral wire, to power a single receptacle. Thus, 240-volt outlets are intended for use specifically with heavier appliances, which need more electricity to operate. For example, a water heater and a central air conditioner require 240-volt wires.

DID YOU KNOW?

(kW), and apparent power is measured in kilovolt amperes (kVA). Thus, the power factor is an expression of energy efficiency and is a dimensionless number in the closed interval of –1 to 1.

What are the differences between single-phase power and three-phase power?

In terms of electricity, the distribution of load is given by phases, single-phase and three-phase power, respectively. AC power also alternates cyclically, which means that first it flows in one direction in a circuit and then·the flow is reversed to the other direction.

- Single-phase power consists of two wires. Typically, one is the conductor wire (power wire/phase wire), and the other is the neutral wire. The conductor is responsible for the current flowing through the load. This produces a single, low voltage.
- Three-phase power provides three alternative currents in the circuit and consists of four wires: three conductors and one neutral wire, respectively. Each of the three wires is 120 degrees apart.

Commercial places require a three-phase supply to use higher loads, whereas a single-phase power supply usually serves households. A three-phase power supply delivers power at a more efficient, steady, and constant rate than a single-phase power supply. Three-phase power supplies have three or four wires and use less conductor material to transmit a set amount of electrical power than single-phase power supplies. Thus, the costs of cabling and total installation is less compared to the power received.

What is a digital multimeter?

A digital multimeter is an electronic measurement tool used to measure two or more electrical values: voltage, current, resistance, capacitance, temperature, or frequency. It is a standard diagnostic tool for technicians in the electrical/electronic industries. These incorporate the testing features of single-task meters: the voltmeter (for measuring volts), ammeter (amps), and ohmmeter (ohms).

What is resistance?

Resistance is the opposition to the flow of current in a circuit. Resistance is measured in ohms, named after German physicist George Simon Ohm and denoted by the Greek letter omega (Ω). Resistance tends to vary with the composition of the material (conductors or inductors), temperature, and shape. For any given conductor, resistance increases with length and decreases with cross-sectional area. The opposite of resistance is conductance, which is how effortlessly an electrical current passes through a conductor.

What is Ohm's law?

In 1827, George Simon Ohm noticed the unique relationship between voltage, current, and resistance. Through a series of tests, he determined that

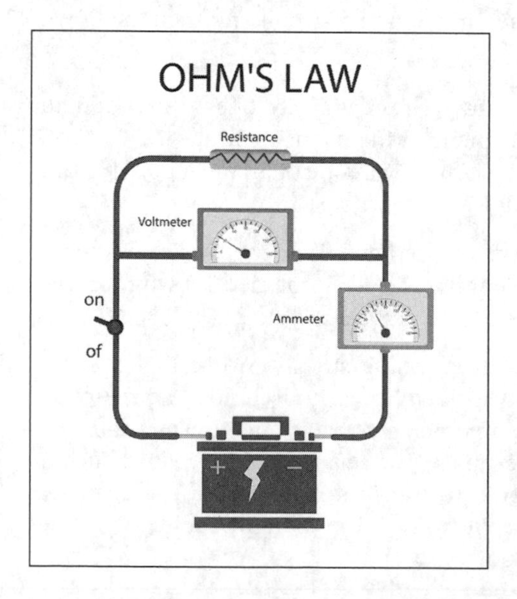

Figure 2: Ohm's Law describes how an electric current is related to voltage and resistance in a circuit.

"the current through a circuit is directly proportional to the voltage applied at the terminals and inversely proportional to the resistance." When the voltage in a circuit increases, the current increases proportionally, providing that the resistance is kept constant. Alternately, when the resistance in a circuit is increased, the current decreases, providing that the voltage is kept constant. The following formula represents this relationship:

V (voltage) = I (current) X R (resistance)

What is power?

Power is the rate at which work is done. Power is measured in watts, denoted by W and named after Scottish mechanical engineer James Watt. The watt is defined as a current of 1 ampere passing through a resistor across a voltage drop of 1 volt, thereby producing 1 watt. Power converts electrical energy to other forms of energy such as heat, light, mechanical, sound, etc.

What is a kWh (kilowatt hour)?

A kilowatt hour is a unit of energy equivalent to 3600 kilojoules. Electric utilities generally use it as a billing unit for electricity provided to customers.

What is electrical energy?

Electrical energy is the energy absorbed by an electrical circuit. This energy has been converted from another form of energy, whether potential or kinetic. The principle of conservation of energy states that energy cannot be created or destroyed; instead, it is converted from one form to another.

HISTORY

How did electrical engineering get started?

Thomas Edison opened the first commercial power plant in the United States in New York in September 1882. While this was not the first breakthrough made regarding electricity, the power plant's opening was particularly significant. At the time of its first operation, the Edison plant served 59 customers. In

Famous American inventor Thomas Edison was also responsible for opening the first commercial power plant in the United States, which he did in 1882.

that same year, the Massachusetts Institute of Technology established the early electrical engineering curriculum in the United States. However, the term "electrical engineering" was first used in 1881 when the group now known as the Institute of Electrical and Electronics Engineers (IEEE) produced documentation that named a new branch of engineering. Initially, practitioners, i.e., those who engaged in electric lighting, were known as electricians.

Before Edison's power plant and the establishment of an electrical engineering title and curriculum, other electric-related phenomena made significant strides in understanding how electricity works and what we could do with it. For example, in 1866, the first successful telegraph cable was operated, followed by the first working telephone in 1876. Three years later, in 1879, Edison invented and used his first light bulb.

Who are the pioneers of electrical engineering?

Some of the more prominent pioneers in electrical engineering include:

Electrical Engineering Pioneer	Innovation
Benjamin Franklin	Lightning rod
Nikola Tesla	Alternating-current (AC) induction motor

Benjamin Franklin was known for being both a statesman and inventor. Along with such items as bifocal eyeglasses and the pot-bellied stove, he invented the lightning rod.

Electrical Engineering Pioneer	Innovation
Thomas Murray	Redesigned early power plants, fuse box
Thomas Edison	Electric light
Charles Brush	Arc lamps for street lighting
Guglielmo Marconi	Radio
Lee de Forest	Audion vacuum tube
Philo T. Farnsworth	Television
George Westinghouse	Improved the power grid
Charles Proteus Steinmetz	Alternating currency
Claude E. Shannon	Father of Information Theory
William Stanley, Jr.	Power transformer
Jack Kilby	Integrated circuit
Frank Sprague	Electric streetcar
John Bardeen	Transistor effect and other major advances
Zénobe Théophile Gramme	DC dynamo
Harry Nyquist	Sampling theorem, supporting digital encoding of analog signals

ELECTRICAL DEVICES AND MATERIALS

What is an electric circuit?

A simple electric circuit consists of four essential components: source, control device, load, and conductor. The source provides the voltage, which could be a battery; the control device operates the circuit and maybe includes a switch; the load consumes the power, perhaps a lamp; and the conductor provides a path, such as a wire. Figure 3 illustrates a simple circuit. For more complex circuits, multiple sources, control devices, loads, and conductors may be present.

Electric circuits can be one of three connections: series, parallel, and series-parallel. A series circuit is one in which all loads are connected so that the end of one load is directly connected to the other. In a series circuit (illustrated in Figure 4), all loads share the voltage while the current remains the same, as the current only has one path to flow through. The total resistance is the sum of all resistances in the circuit, i.e., total resistance = resistance 1 + resistance 2 + resistance 3 … resistance n. Similarly, the total voltage is the sum of the voltage across each component, i.e., total voltage = voltage 1 + voltage 2 + voltage 3 … voltage n.

In a parallel circuit (illustrated in Figure 2), the first end of all loads is connected and the second end of all loads is connected. The current is shared among all loads because the current has multiple paths to flow through while the voltage remains constant, as all loads are connected directly across the source. The total resistance is the inverse sum of all resistance in the circuit, i.e., 1/total resistance = 1/resistance 1 + 1/resistance 2 + 1/resistance 3 … 1/resistance n. In a series

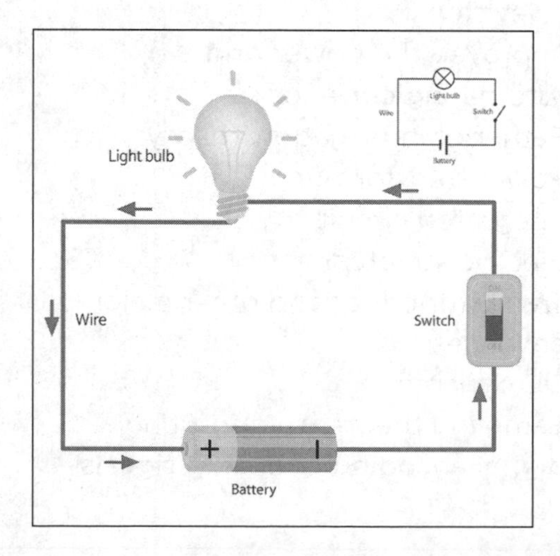

Figure 3: A simple circuit.

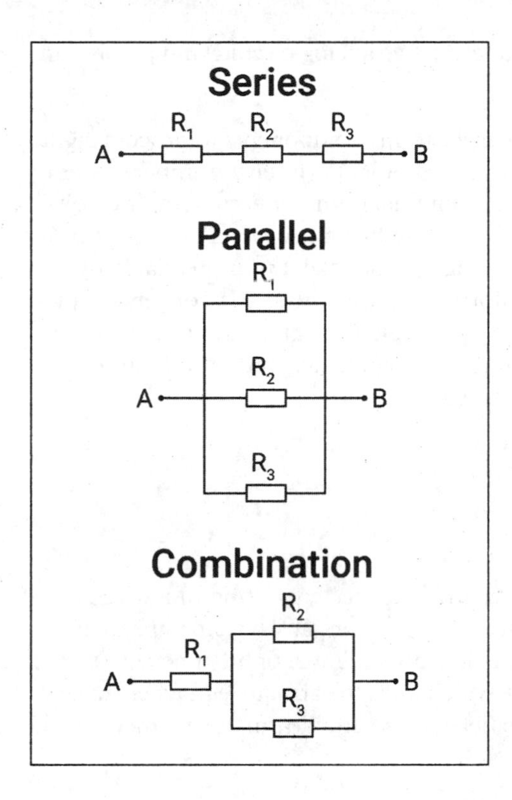

Figure 4: Series, parallel, and series-parallel (or combination) circuits.

circuit, the total resistance is always greater than the value of any one resistor. In contrast, in a parallel circuit, the total resistance is always less than any one resistor's value. The total current is the sum of all currents through the loads, i.e., total current = current 1 + current 2 + current 3 … current n.

A series-parallel circuit (illustrated in Figure 2) contains a combination of loads connected in series and parallel with each other.

What is the difference between analog and digital circuits?

An analog circuit is a type of electronic circuit that uses analog components such as resistors, capacitors, diodes, transistors, etc., to process continuous-valued signals (analog data). These circuits can transform the original signal into another format, such as a digital signal, and can modify signals by adding noise or distortion. This does not require conversion of input signals before processing, i.e., the circuit directly performs various logical operations on analog input and generates an analog output. Hence, no information loss occurs. In most cases, analog circuits are custom made and lack versatility. Analog circuits are classified into two types, namely active analog circuits and passive analog circuits.

A digital circuit consists of digital electronic components to process digital signals with one of two distinct levels, i.e., 0s and 1s (binary number system). These are simple to design and require minimal human interaction. It involves converting input signals to digital form before being processed, then output is again converted from digital to analog signals to make it more readable by humans. Therefore, it contributes to information loss due to the conversion process on the input and output side (analog to digital). Digital circuits at low levels consist of a combination of transistors and logic gates and, at a high level, it consists of microcontrollers and processors.

What is a solenoid?

A solenoid is a kind of electromagnetic device consisting of a wire coil, a housing, and an armature. To produce linear motion, this system aims to create a regulated magnetic field through a coil wrapped into a tightly packed helix. A magnetic field forms around the coil, which pulls the armature in when an electrical current is applied. More specifically, a solenoid transforms mechanical work into electrical energy.

What is a transistor?

A transistor is a semiconductor device with three connections. In addition to rectification, it is used for amplifying or switching electronic signals and electrical control. A voltage or current applied to one pair of the transistor's terminals regulates the current through another pair of terminals. A transistor can amplify a signal because the controlled (output) power can be higher than the controlling (input) power.

What are cells and batteries?

In DC circuits, the voltage is often provided by a constant voltage source such as a battery. The battery converts chemical energy to electrical energy. A battery consists of two or more electrochemical cells connected in a series, parallel, or series-parallel.

A basic chemical cell consists of two lead connectors with a connecting material, known as the electrolyte. The positive lead or side is known as the cathode, while the negative lead or side is known as the anode. When a load is connected across the leads, the resulting chemical reaction between the electrolyte and the leads creates a current that flows from one lead, through the electrolyte, to the other. As the cell discharges across the load, the leads either deteriorate or the electrolyte dries up over time, and the cell is unable to produce any more current.

What is a conductor?

A conductor is a material that has many free electrons that continually move from one atom to the other. In electric circuits, it is necessary to use conductors to connect circuit components that have low resistance. Some common examples of good electrical conductors are copper and aluminum. Other conductors can be used in electrical circuits, such as gold, silver, and platinum, but these are often very expensive and therefore not feasible to be used for circuits. Water is also an excellent conductor of electricity.

What are the properties of a good conductor?

- Low or zero resistance for the flow of electricity
- Good mechanical strength
- Less specific weight
- High corrosion resistance
- Infinite material conductivity
- Charges occurring only on its surface
- Less resistance variation with temperature
- All points at the same potential
- High ductility

What factors do a conductor's resistance depend on?

- Cross-sectional area
- Length
- Resistivity
- Temperature

- Nature of the current flowing through it
- Thickness of resistor

What is an RLC circuit?

An electrical circuit consisting of a resistor (R), a capacitor (C), and an inductor (L) that are connected in a parallel or series is termed an RLC circuit. The voltage in the capacitor lets the current flow stop and flow in the reverse direction. This results in an oscillation or resonance. Thus, an RLC circuit is also known as an oscillating circuit and a second-order circuit.

What is Norton's theorem?

Norton's theorem defines that any linear circuit consisting of an independent or dependent voltage source, current sources, and different circuit compo-

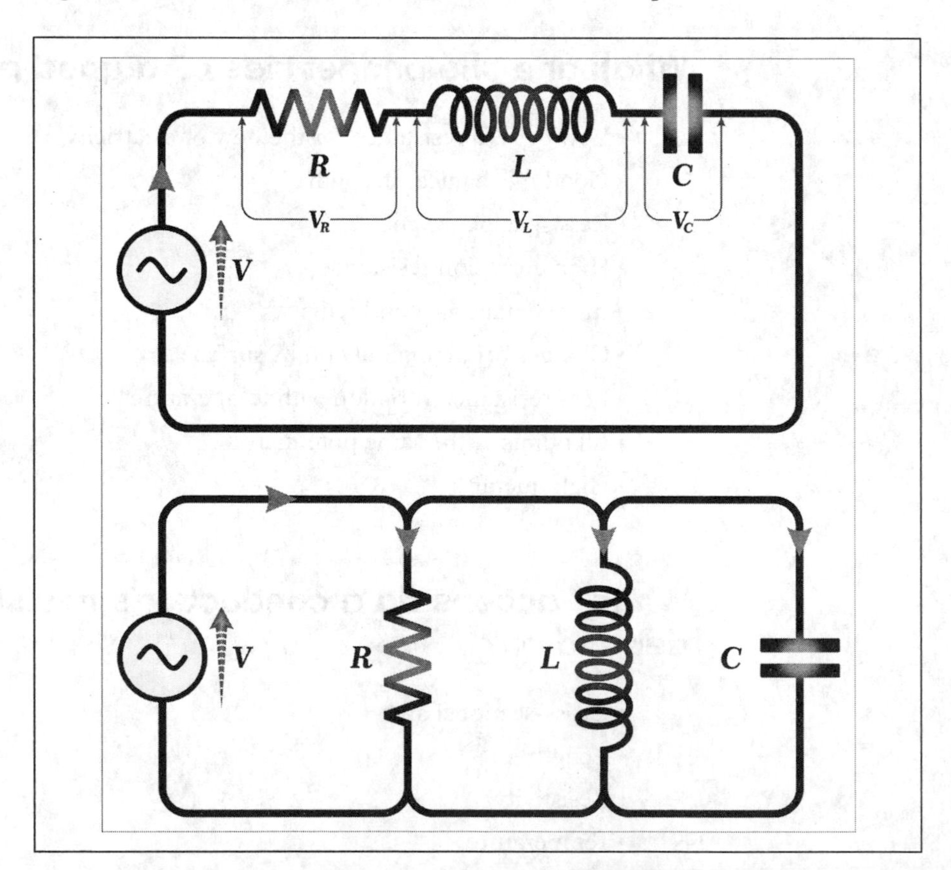

Figure 5: Series (top) and parallel (bottom) RLC circuits.

nents (no matter how complex) can be substituted by an equivalent circuit consisting of just a single current source and parallel resistance linked to a load.

What is an insulator?

An insulator is a material that has only a few free electrons. In insulators, electrons are tightly bound by the nucleus and are therefore not easily moved from one atom to the next. By this characteristic, insulators have very high resistance and cannot connect circuit components.

What are the properties of a good insulator?

- Large dielectric strength
- High air permeability
- High resistivity
- Nonporous and free from impurities
- No effect by temperature
- High mechanical strength
- Nonhygroscopic
- High resistance to corrosion
- High breakdown voltage

What are semiconductors?

Semiconductors are materials (class of crystalline solids) that have an electrical conductivity value falling between conductors (general metals such as copper) and insulators/nonconductors (such as glass or ceramics). Semiconductors can be pure elements (such as silicon) or compounds (such as gallium arsenide). Their conductivity may be altered by doping, i.e., adding small quantities of impurities to the crystal structure (pure semiconductors).

Semiconductors fall into two broad groups:

- Intrinsic semiconductors: These are incredibly pure. They are composed of only one kind of material: silicon (Si) or germanium (Ge). These are often referred to as "undoped semiconductors."
- Extrinsic semiconductors: These consist of some other impurities (substances) that changed their properties, i.e., they have been doped with another material.

DID YOU KNOW?

What are Kirchhoff's laws?

Kirchhoff's current law (first law) states that for any node (junction) in an electrical circuit, the sum of currents flowing into that node is equal to the sum of currents flowing out of that node; i.e., the algebraic sum of the electric currents that intersect at a point is zero in any network of wires.

Kirchhoff's voltage law (second law) states that the directed sum of the potential differences (voltages) around any closed loop is zero.

What are circuit conditions?

Electrical circuits are always designed to ensure that the current path is continuous and that conductors are properly insulated. However, sometimes, conductors lose their insulation or the circuit component becomes defective or is removed from the circuit. Two conditions can happen within a circuit to pose safety concerns:

A short circuit happens when a direct connection between two conductors of opposing polarity occurs that bypasses the load. In a short circuit, the full system voltage in one conductor connects directly with another conductor on the circuit's return side. Because this connection bypasses the load, no current is present in a short circuit, and the resistance is meager.

An open circuit occurs when the continuity between components is broken and the path for the current to flow is no longer complete. In an open circuit, the full system voltage applied can be measured across the open terminals, but no current is present.

What are overcurrent protective devices?

Excess current in any circuit can be hazardous. A short circuit or circuit overload can cause this excess current. To protect against the harmful effects of overcurrent, such as fires or other property or self-damage, it is necessary to use devices designed to detect overcurrent and interrupt the circuit before any harm is caused. The two main types of overcurrent devices are fuses and circuit breakers.

What is an automatic voltage regulator (AVR)?

An automatic voltage regulator (AVR) is an electronic device used in generators to automatically maintain and regulate constant voltage levels, ensuring that fluctuating voltage levels are transformed to provide constant voltage levels and a stable supply of electricity. Thus, it also prevents electrical equipment from sustaining damage caused by surges/fluctuations in voltage.

What are fuses?

Fuses are overcurrent protective devices that are designed to protect the electrical circuit from overcurrent. Fuses are designed with specific voltage and current ratings that will break the circuit when the current circuit is frequently above its rated value. The fuse consists of a strip of wire, called the fuse element, that completes the path between the top and bottom of the fuse. The fuse is always connected in series with the rest of the circuit to ensure that the full circuit current continually passes through the device. The fuse element is often made of a material that can withstand the heat created by the flow of current at its rated value. The material is always sensitive enough to withstand occasional current surges. However, when the current constantly flowing exceeds the circuit's rated value, the fuse element melts and thereby safely interrupts the circuit.

What are circuit breakers?

Circuit breakers, like fuses, are designed to protect electrical circuits from sustained excess current. Circuit breakers are switches that operate automatically to break the circuit when an overload or short circuit causes the current to rise and remain higher than the expected value for an extended period. Circuit breakers are rated for voltage and current and are often used in residential, commercial, and industrial buildings. The circuit breaker is connected in series with the circuit between the source and the control device, such as a switch. The most

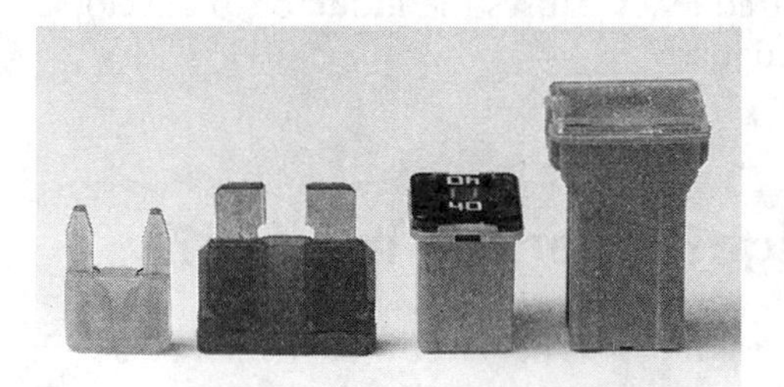

Blade and cartridge fuses like these are commonly found in automobiles.

common circuit breakers are basic switches that have two contacts connected using a bimetallic strip. In its normal operation conditions, the contacts are closed when an excess current passes through the breaker for a sustained period of time. The bimetallic strip moves due to the increased heat caused by the current. When the strip moves, it breaks the connection between the switch contacts and successfully interrupts the circuit. Unlike the fuse, the circuit breaker is most often used because it is easily reset after the fault is detected and removed from the circuitry. Fuses, on the other hand, must be completely replaced.

What are capacitors?

Capacitors are electrical circuit components that are used to store electric charge. A capacitor consists of two conductive plates separated by some insulating material. The effect of the capacitor in the circuit is called capacitance and is measured in farads. The farad is named after British physicist Michael Faraday. The farad is determined as the value of 1 farad of capacitance developed when 1 volt is applied across the plate of the capacitor to move 1 coulomb of electrons from one plate to the next. When the circuit is connected, the capacitor's charge tends to build up as current flows through the device. When the capacitor is fully charged, it dissipates its charge through the load in the circuit.

What are inductors?

Inductors are circuit components that are used for several reasons. Inductors are called passive electrical components because they oppose sudden changes in the current moving through the circuit. When the current in a circuit suddenly increases due to surges in voltage from the source, the inductor temporarily stores the excess current in an electromagnetic field and slowly releases it back into the circuit. The inductor consists of a coil of wire, often made from copper, wound around a core that can be made from air, ceramic, iron, or ferrite. When connected within a DC source circuit, the inductor acts like a wire piece that connects the circuit. This means that the inductor has little to no effect on the overall resistance in the circuit. However, when connected within an AC source circuit, the source's frequency causes the inductor to develop a resistance value known as the impedance.

What is a Faraday cage? Where is it used?

A Faraday cage is a structure used to obstruct electromagnetic fields with a continuous covering of conductive material. It creates a barrier between the de-

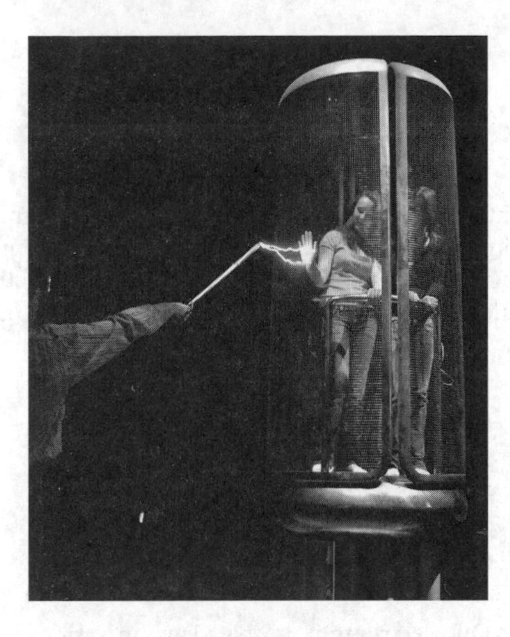

A Faraday cage is demonstrated at the Palais de la Découverte in Paris.

vice's internal components and the external electric fields. It operates when an outside electric field allows the electric charges to be redistributed within the container's conducting material, which in turn cancels the influence of the field within the cage. Various applications include microwaves, protective gear for electricians and linemen, lightning safety, MRIs (magnetic resonance imaging), etc.

What is a transformer? How does it work?

A transformer is a passive electrical device that transfers electrical energy from one circuit to another through the method of electromagnetic induction. It is most often used between circuits to increase or decrease voltage levels.

A varying current creates a variable magnetic flux in the transformer's center in any one coil of the transformer, which produces a variable electromotive force in any other coils wound around the same core. Thus, mutual induction between two or more coils facilitates the flow of electrical energy between circuits.

The transformer working principle is based on Faraday's law of electromagnetic induction. When an alternating electrical source is supplied to a coil, the alternating current produces a continually changing and alternating flux surrounding the winding. Thus, when another coil is brought close to it, some portion of this alternating flux will link with the second coil, and an EMF will be induced in the second coil. If the circuit of this secondary winding is closed, then a current will flow through it.

DID YOU KNOW?

Where does electromagnetism come from?

When an alternating current is passed through a piece of wire, the current fluctuating value creates a pulsating, magnetic field. This is known as electromagnetism. The field's magnitude is highly dependent on the rated value of the current, the length and cross-sectional area of the wire, and the distance of the wire from which the field is being measured. The direction of the current determines the direction of rotation of the field.

What is a stepper motor? What are its uses?

Stepper motors are brushless, DC electric motors that drive in various (discrete) steps and divide a full rotation into several equal steps. They have several coils in groups called "phases." The motor would spin, one stage at a time, by energizing each process in series. We can achieve very precise positioning and/or velocity control with a computer-controlled method. Thus, stepper motors are used in precision motion-control applications. Many applications of stepper motors are found particularly in areas where precise positioning with a motor is required; for example—hard disk drives, antennas, robotics, telescopes, etc.

What is a diode? Where is it used?

A diode is a semiconductor device that functions as a one-way switch. It allows electrical current to flow freely in one direction but not in the other. This two-terminal electronic component works with the help of a built-in electric field. The diode's polarity is defined by an anode (positive lead) and a cathode (negative lead). A diode is said to be forward-biased when it allows current flow. In contrast, it is called reverse-biased when the diode acts as an insulator and does not permit current to flow.

A diode has many critical uses in modern technology; a blocking diode is used in some circuits for load protection against the accidental reverse current used for rectification of AC power to DC power. Also, it is used as a voltage regulator to set the input voltage at the desired level and avoid any fluctuations in the supply.

What is reverse polarity? Why is it dangerous?

Reverse polarity is when hot and neutral connections are wired "backward/incorrectly" at a receptacle. The hot wire is electrically energized in normal conditions, usually up to around 120 volts, whereas the neutral wire is 0 volts. Therefore, if the socket's polarity is inverted, it means that the neutral wire is connected to where the hot wire is intended.

In such cases, it may produce a shock risk even when an appliance switch is open. As 120 volts are now connected to the wire that would typically be the neutral wire, making it still energized due to reverse polarity, this makes it dangerous and leads to a shock hazard.

What is a relay? How does it work?

Relays are electromagnetic switches that use electromagnetism to open and close circuits. They are controlled by a comparatively small electrical current and transform small electrical impulses into larger currents. Hence, by opening and closing contacts in another circuit, relays regulate one electrical circuit.

How do car alternators work?

An alternator is an electrical generator that transforms mechanical energy into electrical energy in an alternating current. In modern vehicles, it is used to

Semiconductor diodes on a circuit board.

charge the battery and control the electrical device while running. An alternator is a significant part of a vehicle's charging mechanism. These are present in every car with internal combustion except for certain hybrids. The alternator charges the battery and provides extra power while the engine is running.

The alternator's main components include:

- Rotor and stator: When the car is in motion, a series of magnets on the rotor spin fast within the stator, thus generating an alternating current.
- Diode: This transforms the alternating current into a direct current used to charge the battery of the car and ensures that the electric current only passes in one direction: from the alternator to the car battery.
- Voltage regulator: This ensures that the voltage remains within an acceptable range throughout the charging device.
- Cooling fan: This helps to dissipate the heat.

How does a dynamo work?

A dynamo is an electrical generator that creates a direct current. It consists of three main components: the stator, the armature, and the commutator. In this, the electromagnets are stationary and are called field coils. The current is produced in an armature. As the armature turns, the current changes direction. The commutator is a rotary switch that disconnects the power during the reverse-current part of the cycle, so the current coming out always flows in the same direction.

What are the advantages of using alternators over dynamos?

- An alternator is a three-phase unit as opposed to a single-phase unit.
- An alternator will give a much higher charge rate at slow speeds (idle).
- The voltage regulator on an alternator is less complicated than it is on a DC generator.
- Alternators are cheaper, smaller, and more durable.
- At idle speed, alternators can provide useful charging.
- Alternators use slip rings, which increase brush life substantially more than a commutator.

DID YOU KNOW?

What is a light-emitting diode (LED)?

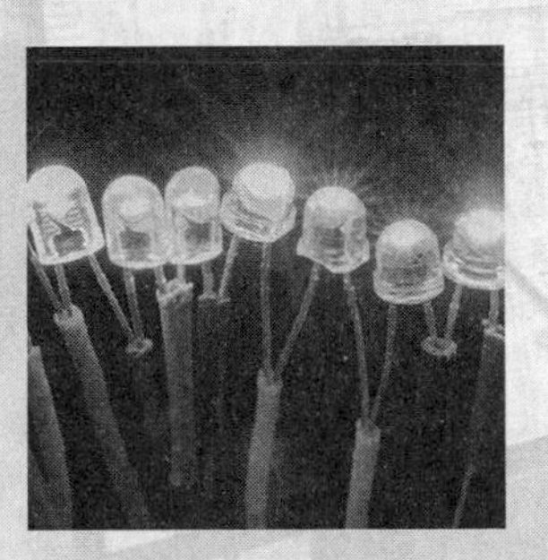

A light-emitting diode (LED) is a semiconductor device (light source) that emits light when an electric current passes through it. Electrons in the semiconductor recombine with electron holes, producing energy in the form of photons. Thus, as the particles that hold the current interact together inside the semiconductor material, light is emitted. The light color is affected by the energy needed for electrons to cross the semiconductor's band gap. LEDs are defined as solid-state devices because the light is produced inside the solid semiconductor material.

How does an induction motor work?

It operates on the principle of induction. By applying a three-phase current through the rotating electromagnetic field of the stator winding, an electrical current in the rotor is produced. This generates a current within the rotor's conductors and creates the rotor's magnetic field, which attempts to obey the stator's magnetic field, forcing the rotor into rotation. The rotor of an induction motor may either be a wound type or a squirrel-cage type. AC induction motors are asynchronous machines, which means that the rotor does not spin at the same speed as the stator's rotating magnetic field. To produce the induction into the rotor, a specific variation in the rotor and stator speed is required. The variation gap is called the slip.

What are three-way and four-way switches?

Three-way switches have three terminals (besides a ground terminal) and make it possible to control one electrical fixture from two different locations. For example, adding three-way switches at both ends of a corridor or vast space helps one to turn the light fixture on or off from both positions.

A four-way switch is similar to a three-way switch. It has four terminals and a ground terminal. It helps control one fixture from three locations. They are connected to form a circuit of two three-way switches.

DID YOU KNOW?

How does a car battery work?

Usually, car batteries depend on a lead–acid chemical reaction. Such batteries are termed SLI, which stands for "starting, lighting, and ignition." This type of battery gives short bursts of energy to power the vehicle's lights, accessories, and engine. Thus, it provides the initial current to the starting and ignition systems. Once the battery turns the engine on, the alternator supplies power to the car. Also, the battery provides the current to the other electrical equipment all the time, while the generator/alternator does not charge (engine is stopped).

What are ground, neutral, and hot wires?

A hot wire is used to supply the initial power input to the circuit. It takes its current from the power source to the outlet and offers 120-volt, alternating-current sources. Hot wire is distinguished by its black covering/casing. Additionally, other hot wires can be red, blue, or yellow. However, these colors can represent a different purpose in addition to powering an outlet.

A hot wire initiates the start of a circuit, and a neutral wire completes it. A neutral wire is attached to the ground. It gives the return path for the current that the hot wire provides. Thus, it delivers the circuit back to the original power source, i.e., to a ground attached to the electrical panel. Also, this prevents excess current from residing in the outlet. Typically, white or gray wire is neutral wire.

A ground wire is connected to the ground either directly or through another grounded conductor. It does not hold any current in standard circuit conditions but acts as a safety feature, carries the faulty current away from the electrical system, and directs it toward the ground in case of electrical accidents.

What is an inverter, and how does it work?

Two different electrical power forms are available, namely direct current (DC), which is primarily used by small, digital goods with circuit boards and is provided by batteries and solar panels, and alternating current (AC), which is supplied from the power sockets in homes and is usually used to power larger appliances.

An inverter is effectively a converter of electric power that transforms direct current (DC) into alternating current (AC). The primary use of an inverter is to provide backup power when a flow of electricity is not available. The most common examples include a house inverter, which supplies 110/220V alternating current to the house when the power goes out, and a power inverter, which turns the energy stored in a battery into alternative power for industrial and commercial applications where no power is available.

What are time-delay relays? What are their applications?

Relays are important electrical components because they can be used in a circuit to switch between high and low current. This occurs through electromagnetism. In a standard-control relay, connections close instantly when voltage is supplied to the coil and open immediately when the voltage is withdrawn. In many scenarios/applications, it is beneficial to get the operation of the connections to be postponed/delayed after supply or withdrawal of voltage occurs. TDR is simple, efficient, and provides economical control.

Thus, time-delay relays are useful to control the electrical power flow when devices must be powered for a set time period or if the device must be de-energized after a set time period. Some common applications include dishwashers, roller coasters, fans that cool off a warm motor, or to power off the light in the garage door opener after a set time period.

What is a PLC (programmable logic controller), and how does it work?

A PLC (programmable logic controller) is a solid-state digital computer used for industrial automation to monitor input and output devices and automate different electromechanical processes to make logic-based decisions. Thus, it makes decisions based on a customized program to control the state of output devices. PLCs are versatile and durable control systems that can be modified to work with almost any application to eliminate problems such as high power consumption, etc. A PLC setup includes a CPU, analog inputs, analog outputs, and DC outputs. Its design works with inputs, outputs, a power supply, and external programming devices.

Four steps that take place in PLC operations are:

1. Input scan: Identifies the status of all input-connected devices to the PLC

2. Program scan: Runs program logic generated by the user
3. Output scan: Energizes or deenergizes all PLC-connected output components
4. Housekeeping: Communicates with programming terminals, internal diagnostics, etc.

These steps take place in a repeating loop in a PLC.

What are the various types of transmission cables based on thermal capacity?

Cables are divided into three categories:

1. Low-voltage cables: These can carry up to 1,000 volts.
2. High-voltage cables: These can carry up to 23,000 volts.
3. Super-tension cables: These can carry between 66,000 to 132,000 volts.

What is the difference between electrical circuits and electronic circuits?

The key distinction between electrical and electronic circuits is that electronic circuits have decision-making capability, while electrical circuits do not have this. An electric circuit only powers machines with electricity. However, an electronic circuit can interpret a signal or command and perform a task to fit the need. For instance, a microwave oven makes a sound (beeps) when it has finished cooking to remind the user that the meal is ready.

A programmable logic controller (PLC) monitors input and output devices and automates electromechanical processes. It is designed to operate under harsh industrial conditions such as automobile plants and other manufacturing facilities.

CAREERS

What is the future scope of electrical engineering?

Electrical engineering is a fascinating sector wherein one understands how the technology around us functions. This field has excellent tools and building blocks to design modern instruments and overcome complicated problems. It is perhaps one of the most critical divisions of engineering, with complex topics such as control systems, power engineering, instrumentation, etc. Electrical and electronics engineers work in research and development, manufacturing, management, telecommunications, and the federal government. According to the U.S. Bureau of Labor Statistics, the employment of electrical and electronics engineers is expected to rise by 3 percent, almost as high as the rate for all occupations.

What are some career pathways to electrical engineering?

Computer Engineering

- Creating and manufacturing computer hardware and smart applications
- Diving deep into artificial intelligence, cybersecurity, and software development

Biomedical Engineering:

- Developing human-made products and materials
- Designing modern medicines, surgical implants, genetically modified species, and medical equipment

Engineering Management:

- Combining engineering knowledge with teamwork and project-management skills
- Managing large teams and addressing issues

Engineering Communications:

- Handling information collection, transmission, and storage with finite capacity systems
- Examining and processing EKG patterns, sound signals, pictures, and videos
- Evaluating advanced biometrics and engineering optimization strategies
- Working in the fields of telecommunications and robotics

What certifications are available for electrical engineering?

The most common prominent certifications include:

- AutoCAD (electrical)
- PLC programming and operation certification
- SCADA (supervisory control and data acquisition)
- Cloud Computing
- Embedded and VLSI Systems
- Robotics and Intelligent Systems
- Power Electronics
- Signal Processing
- Hardware Networking
- Licensed Professional Engineer
- Industrial Automation
- PMP certification

What are some operational challenges for electrical engineers?

- Trying to adopt renewable resources to reach net-zero carbon pollution
- Fulfilling EV market requirements; substantial infrastructure investment
- The need for introducing charging networks
- Using technologies like harmonic current filters to regulate the effect of harmonics
- The reliability of the machinery being integral to maintaining good power quality
- Electrostatic discharge issues
- Electrical equipment protection
- Issues of product performance and reliability
- Problems related to outsourcing
- Staying updated with modern technologies

What are the subdisciplines of electrical engineering?

Electrical engineering consists of many areas of specialization. In most cases, electrical engineers work in one or more subdisciplines or know more than one of these subdisciplines. The most common subdisciplines are power, control, electronics, microelectronics, signal processing, telecommunications, instrumentation, and computer.

- Power engineering: Electrical engineers who specialize in power focus primarily on the generation, transmission, and distribution of electricity in various sectors. Power engineers ensure that the power grid can provide and maintain the power needed to support all the power systems in a particular region or area.
- Control engineering: Control engineers focus on maintaining and modeling control systems to fulfill a variety of needs. For example, control engineers use programmable-logic controllers (PLCs) to monitor the manufacturing process, motor functions, and other real-life application systems.
- Electronics engineering: This type of engineer often designs and creates electronic circuits of varying magnitudes. From the circuitry of a cell phone to a drone or satellite, electronic engineers work with computers, microprocessors, and radio circuits to design complex electrical circuitry.

Power engineers specialize in the generation, transmission, and distribution of electricity.

- Microelectronics engineering: Microelectronics engineers focus mainly on the study, design, and manufacture of semiconductor devices; this subfield of electrical engineering deals primarily with micrometer-scaled devices.
- Signal-processing engineering: An engineer working in this subdiscipline manipulates signals through analysis, synthesis, and modification strategies to decipher any information embedded in the signal. These engineers work on projects that often include other subdisciplines, such as telecommunications, audio, and broadcast engineering, among others.
- Telecommunications engineering: This subdiscipline involves the use of communication channels to transmit information across large distances. These electrical engineers work to enhance and maintain communication systems that often range from simple circuitry to more strategic and robust designs.
- Instrumentation engineering: Instrumentation engineers center on designing, testing, and evaluating electrical instruments used to measure various physical quantities, such as pressure, temperature, voltage, smoke, etc. Instrumentation techniques and tools span the breadth of human existence from pressure systems in airplanes to sensors in automobiles and even smoke detectors in homes.
- Computer engineering: This subdiscipline combines several computer science areas with electronic engineering to produce engineers who focus on computer hardware and software design. Computer engineers work on improving and designing new, innovative computer systems that encompass new and existing technologies.

Mechanical Engineering

What is mechanical engineering?

Mechanical engineering is one of the broadest engineering professions. It focuses on making innovations to address human issues with an adequate understanding of core areas—for example, knowledge and application of mechanics, thermodynamics, materials science, etc., to deal with the design, development, and utilization of machines. For this design and analysis, mechanical engineers use computer-aided design (CAD), computer-aided manufacturing (CAM), and product life cycle management. Thus, it includes the design, production, and operation of machinery.

HISTORY

What is the history of mechanical engineering?

With the development of civilization, the emergence of mechanical engineering started from the first steam engine, which is dated from ancient Greece, the work of Hero (or Heron) of Alexandria. Then, at that point, during the Islamic Golden Age (from the seventh to the fifteenth century), Ismail al-Jazari's book *The Book of Knowledge of Ingenious Mechanical Devices* was written in 1206.

The first professional society for mechanical engineers, the Institute of Mechanical Engineers, was formed in the United Kingdom in 1847. The first power-driven machines were made during the nineteenth century; then, during the twentieth century, more advanced systems were built. The development of physics and machine tools during the nineteenth century permitted the separation of mechanical engineering from the rest of the engineering subdisciplines. This led to the manufacturing of machines and engines to power them. Indeed, one of the primary difficulties of mechanical engineering was the creation of power-driven systems. Thus, the understanding of mechanical science, which incorporates dynamics, thermodynamics, power, heat transfer, etc., led to mechanical engineering advancement.

List some famous mechanical engineers and their contributions to society.

The following is a list of pioneers from the field of mechanical engineering:

- James Watt (1736–1819): In 1776, Watt, a Scottish inventor, improved Thomas Newcomen's 1712 steam engine with his Watt steam engine. Watt

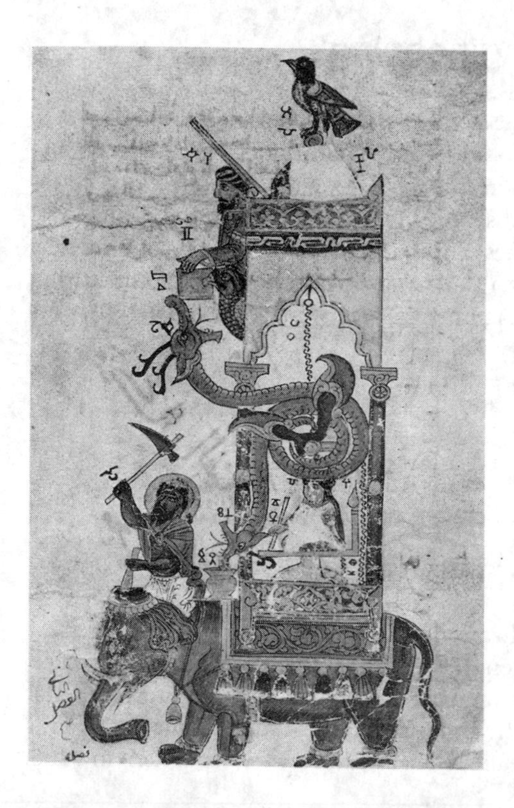

Many of Ismail al-Jazari's inventions were powered by water, such as his famous and elaborate elephant water clock.

created the separate condenser, a design innovation that eliminated energy loss and dramatically increased the power, efficiency, and cost-effectiveness of steam engines. He eventually modified his engine to create rotational motion, considerably expanding its use beyond water pumping.

- George Stephenson (1781–1848): In 1814, Stephenson, an English civil engineer, built the Blücher locomotive, which could move 30 tons of coal up a slope at 4 miles per hour.
- Charles Babbage (1791–1871): Babbage, an English mathematician, was the developer of the first mechanical computer, the difference engine, which served as a precursor to numerous complicated electronic computers. He is recognized as the Father of Computing. The machine's concept inspired many engineers, and the Science Museum in London created one in 2000 using the same design, which was capable of doing mathematical computations up to 31 digits long.
- Rudolf Diesel (1858–1913): Diesel, a German inventor, is well known for developing the first effective diesel engine over a four-year period from 1893 to 1897. After realizing that steam engines were wasting 90 percent of their energy, Diesel focused all of his efforts on developing internal-combustion engines, which used petrol and diesel as fuel. His innovation was eventually used as a replacement for steam piston engines.
- Catherine (Kate) Anselm Gleason (1865–1933): Kate Gleason, an American, was the first woman to enroll in the Mechanical Arts program at Cornell University. She left schools to help out at her father's machine shop. She continued to learn and study engineering even before many programs were open to women. She was the first woman with no bank ties to become president of a national bank. Gleason also designed affordable houses.
- Carl Friedrich Benz (1884–1929): Carl (also known as Karl) Benz was a German automotive inventor and engine designer who invented the first practical vehicle in 1885. After experimenting with self-powered vehicles, Benz developed the Motorwagen, a three-wheeled, petrol-powered carriage that was patented in 1886.

BASIC CONCEPTS

What are Newton's laws of motion? Why are these laws important?

Sir Isaac Newton's first law states that every object will remain at rest or in uniform motion in a straight line unless compelled to change its state by the action of an external force.

Sir Isaac Newton (1642–1726) defined the fundamental laws of motion, gave us our fundamental understanding of gravity, and also is credited with developing calculus (although Gottfried Wilhelm Leibniz is also credited with inventing calculus).

Newton's second law explains how an object's velocity changes when subjected to an external force. The law defines a force equal to a change in momentum (mass times velocity) per change in time.

Newton's third law states that every action (force) in nature has an equal and opposite reaction.

Newton's laws are essential in our day-to-day activities. For example, friction is required between the wheels and the ground in order for a vehicle to move. Since the wheels are spinning, they exert a force on the ground, and the ground, in return, exerts a reaction force on the wheels. This force is responsible for pushing the car forward. Thus, Newton's laws are accountable for many such activities in our routine.

What is the law of conservation of energy?

The law of conservation of energy states that energy can be neither created nor destroyed, but it can be transformed from one form to another.

For example:

- When a moving vehicle hits a parked vehicle and makes the parked vehicle move, energy is transferred from the moving vehicle to the parked vehicle.
- When a player kicks a football that is at rest on the ground in a game of football, energy is transferred from the player's body to the ball, setting it in motion.

What is the first law of inertia?

Sir Isaac Newton's first law states that an object will remain at rest or move at a constant speed in a straight line unless an unbalanced force acts on it.

For example:

- When a car makes a sharp turn, the person sitting inside is forced to move on one side.
- Seat belts get tightened automatically when a car experiences a jerk or stops suddenly.

What are the laws of thermodynamics?

1. The first law, also known as the law of conservation of energy, states that energy cannot be created or destroyed in an isolated system.
2. The second law of thermodynamics says that when energy changes from one form to another, form or matter moves freely and entropy (disorder) in a closed system increases.
3. The third law of thermodynamics is also called Nernst law and states that a system's entropy at absolute zero is a well-defined constant. It also helps to analyze the chemical and phase equilibrium.

What is the meaning of the term "degrees of freedom"?

The minimum number of independent variables required to define positions or different motions in a system is known as the degrees of freedom. In a 3D-spaced system, 6 is the maximum number of degrees of freedom for an unconstrained, rigid body. The number of motions is three translatory movements and three rotary movements along the X, Y, and Z axes. A 2D-spaced system has a maximum of 3 degrees of freedom, with two translatory movements along the X and Y axes and one rotary movement along the Z axis.

What is force?

In science, a force is any interaction leading to push or pull on an object with mass that, when opposed, will change the motion (velocity) of an object.

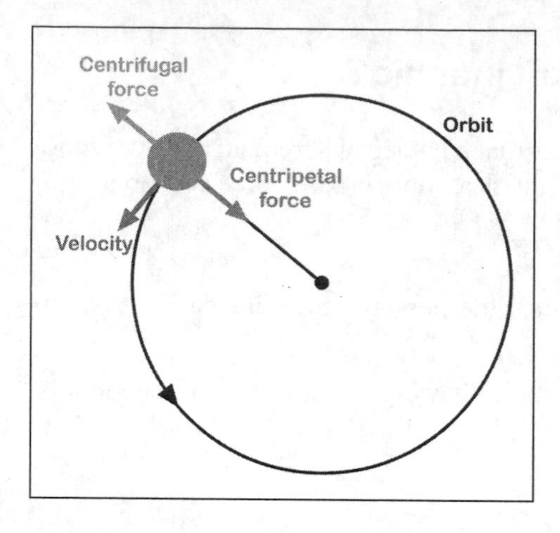

Figure 6: Centrifugal force is the force that acts in an outward direction from an axis, while centripetal force is the force that makes a body follow a curved path by directing it toward the axis.

A force can cause an object to begin moving from a state of rest, i.e., to accelerate. A direction is always associated with the force.

What is centrifugal force?

Centrifugal force is an apparent force that acts outward on a body traveling around a center, arising from the body's inertia. It depends on the object's mass, rotational speed, and distance from the center.

What is work?

Work is force applied over a distance. It is defined as an activity that includes mental or physical effort that causes the movement or displacement of an object to accomplish a result. The standard unit of work is the joule (J). Examples include lifting an object against Earth's gravitation and driving a vehicle up a hill.

What is a pulley? What are its applications?

A pulley is a wheel with a groove along its edge that holds a rope or cable. Generally, at least two pulleys are utilized together. When pulleys are used together along these lines, they lessen the measure of power expected to lift a heap. A crane utilizes pulleys to enable it to lift heavy loads. This acts to alter the direction of a force applied to the rope and raise substantial loads. For example, elevators utilize different pulleys to work, and wells use the pulley system to pull a bucket of water out of a well.

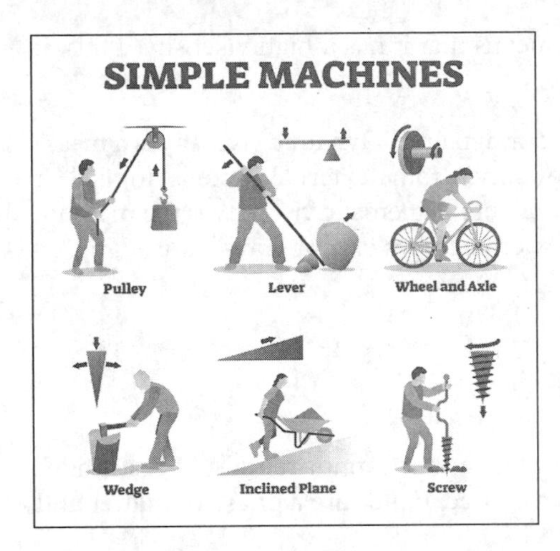

The six basic, simple machines.

What is a lever? What is its main function?

A lever is a rigid rod that rotates around a joint and is used so that a small force can move a much bigger force, giving a mechanical advantage by spreading out the effort over a longer distance and thus reducing the amount of force needed to move/lift an object; examples include fishing rods and ice tongs.

What is the use of a wedge?

A wedge is a triangular-shaped tool that is thicker toward one side and thinner on the opposite end (the edge); it is utilized to separate two objects, split an object, or lift or hold an object in place.

Explain the mechanism behind screws.

A screw is a mechanism that converts rotational motion to linear motion and vice versa. Thus, it is a type of fastener.

Fluids and Viscosity

Define viscosity, its types, and their differences.

In general, viscosity is termed as thickness/resistance to pouring. For example, water is "thin," which means that it flows easily and has a low viscosity, while vegetable oil is "thick," which means that it has a high viscosity. Thus, it is a fluid's internal resistance to flow.

Viscosity can be either kinematic or dynamic. Dynamic viscosity is a measure of a fluid's resistance to shear flow when some external force is applied. It portrays the behavior of fluids under pressure. Kinematic viscosity is the measure of fluid's resistance to flow when no external forces except gravity are acting.

Define fluid and fluid flow.

A fluid can be defined as a substance that cannot remain at rest under the action of any shear stress or external force. Fluids are a phase of matter and include liquids, gases, and plasmas.

Fluid flow is defined as the motion of a fluid subjected to unbalanced forces. Thus, it is the displacement of any quantity from one place to another. It is usually measured in (m^3/hr).

What is Bernoulli's principle? Why is it used?

In fluid dynamics, Bernoulli's principle (named after Swiss mathematician and physicist Daniel Bernoulli [1700–1782]) states that an increase in a fluid's speed co-occurs with a decrease in pressure or a reduction in the fluid's potential energy. For example, since the rate is greater in the narrower pipe, that volume's kinetic energy is more remarkable. Bernoulli's principle is the single rule that clarifies how heavier-than-air objects can fly. It also explains that faster-moving air has low air pressure and slower-moving air has high air pressure. Thus, it describes why plane wings are bent along the top and why boats need to direct away from one another as they pass. The pressure over the wing is lower than the pressure beneath it, giving a lift from underneath the wing.

What is the significance of the Reynolds number?

Named after Osborne Reynolds (1842–1912), the Reynolds number (Re) is a significant, dimensionless quantity in fluid mechanics used to help predict flow patterns in various fluid flow situations. Mathematically, it is characterized as the proportion of inertial forces to viscous forces, which foresees the fluid flow conditions. For instance, it helps in finding out whether a fluid that is flowing has a viscous or nonviscous nature. Thus, the Reynolds number aims to develop the relationship in fluid flow between inertial forces and viscous forces.

Osborne Reynolds after whom the Reynolds number is named, was an expert on fluid dynamics, a quantity that helps predict flow patterns in various situations.

What is the difference between a pipe and a tube?

Pipe—A pipe is a type of vessel (round tubular) used to distribute fluids and gases. It is measured by inner diameter (ID). It can be made from various materials, including ceramic, glass, plastic, and concrete, since it is used to transport gases or liquids, making it essential to know the capacity. The pipe's spherical structure makes it effective at managing friction from the fluid flowing into it.

Tube—A tube is a structural type (round, rectangular, squared, or oval hollow section) measured by outer diameter (OD). It is made from plastic, rubber, or metal alloys such as stainless steel. Since it's used in structural applications, the outer diameter is the most critical measurement.

What is a pump? What is a turbine? What is the difference between a pump and a turbine?

A pump is a device that transfers mechanical energy to fluid. The turbine does the opposite. It transfers the flow energy of fluid to mechanical energy.

How does a turbine work?

A turbine is a rotational mechanical device that extracts energy from a high-pressure fluid flow and converts it into valuable work. In a turbine, the

These turbines inside Hoover Dam are an excellent example of how fluid (in this case, water) power can be turned into mechanical energy, which is then turned into electricity.

energy from a fluid flow turns the three propellers like blades around the rotor and imparts rotational energy; they are connected to the main shaft, which then turns the generator, which produces power. For example, windmills are the primary application of turbines.

Which factors affect the efficiency of pumps?

The efficiency of a pump (centrifugal) depends on the size, speed, and proportions of the impeller and casing.

Stress, Strain, and Malleability

What are stress and strain?

When an external force is applied to an elastic body, deformation occurs in the object's shape and size. Strain is the change in the shape and size of an object due to an externally applied force. Strain is calculated by extension per

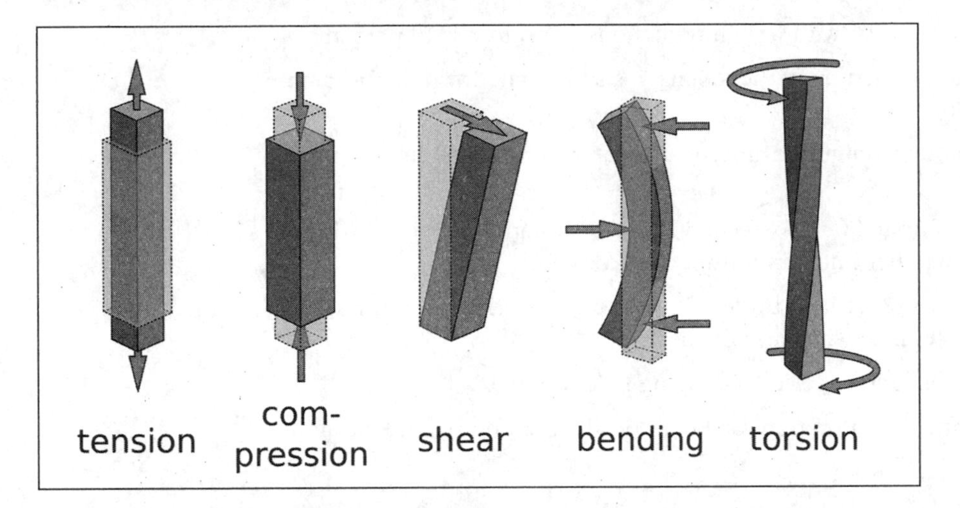

Figure 7: Mechanical stress can be placed on an object in several ways, as this illustration shows.

unit area. Stress, on the other hand, is the internal force associated with strain. Stress is calculated by force per unit area.

What is the difference between stress and pressure?

Stress is related to internal force, which is defined as the reaction produced by the molecules of the body in response to any action that may result in deformation. On the other hand, pressure is an external force, which is defined as force per unit area applied to an object in a direction perpendicular to the surface.

What is surface tension?

Surface tension is a property of a liquid's surface that allows it to withstand an external force due to the water molecules' cohesive existence. Liquid surfaces tend to compress to the smallest possible surface area. Surface tension allows denser items than water to float and slide over a water surface without being partially submerged—for example, razor blades, insects, etc.

What are the benefits of engineering plastics?

New, high-performance composite materials and engineering plastics can provide:

- More functionality, with weight reductions of up to 30–40 percent.
- A significant reduction in the tooling costs associated with high-volume production.
- Software and technique advancements that enhance predictive engineering.
- A savings of up to 30–50 percent on costs by bringing previously distinct elements together into a multifunctional design.
- The ability to resist heat up to 220–230 degrees Celsius and redesign cooling systems to withstand the heat.
- Fast assembly and superior packaging in less space.
- Reduced fuel consumption, pollutant production, noise, and vibration.

What is malleability?

Malleability is a substance's ability to deform under pressure (compressive stress), i.e., it is the capability of being shaped or extended by hammering, forging, etc. For example, lead is a material that is, relatively, malleable but not ductile.

What is ductility? What are some examples of ductile metals?

In materials science, ductility refers to a material's ability to withstand plastic deformation under tensile stress before failure. Some examples of ductile materials include platinum (Pt), copper (Cu), gold (Au), nickel (Ni), and tungsten (W).

Heat and Radiation

Explain conduction, convection, and radiation.

Conduction is the transfer of heat energy by direct contact of bodies. An example is a spoon heating gradually from top to bottom due to direct contact with heated material.

Convection, on the other hand, is the movement of heat by the actual movement of matter. An example is heating water.

Radiation is the transfer of heat energy due to electromagnetic waves or without any medium. An example is solar energy from the sun.

What is a heat exchanger?

A heat exchanger is used to transfer heat between two fluids. When fluid is used to transfer heat, the fluid could be a liquid, such as water or oil, or it could be moving air.

State the difference between nuclear fission and nuclear fusion.

Nuclear fission is the division of one atom into two or more atoms. When a neutron collides with a giant atom, it causes it to excite and split into two smaller atoms, known as fission products. More neutrons are also released, which can start a chain reaction. Since they are simple to initiate and control, uranium and plutonium are the most widely used fission fuels in nuclear power plants. Nuclear fission releases heat by splitting atoms. A tremendous amount of energy is released as each atom splits. The energy released by fission in these reactors heats water into steam. The steam is used to spin a turbine to produce carbon-free electricity.

Nuclear fusion combines lighter atoms into one heavier atom, such as when two hydrogen atoms merge to form one helium atom. It is the same mechanism that drives the sun and generates vast quantities of energy many times

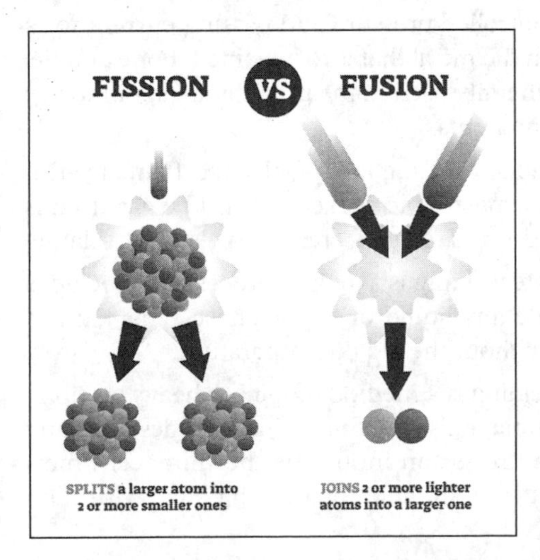

Figure 8: In nuclear fission, atoms are split apart, releasing tremendous amounts of energy; in nuclear fusion, atoms are combined, which also releases energy. Fission creates a lot of radioactive waste, but fusion requires so much initial energy that it has not yet been achieved by humans.

that of fission. It also doesn't contain heavily radioactive fission materials. These are impossible to maintain for long periods of time due to the enormous pressure and temperature needed to join the nuclei together.

Welding

What is welding?

Welding is a fabrication technique for joining materials, typically metals or thermoplastics, by melting them together and causing them to cool, resulting in fusion.

A parent material refers to the parts that are joined together. Filler or consumable material is the substance used to help form the join. Consumables are generally chosen to be identical in composition to the parent material, resulting in a homogeneous weld. Sometimes, a filler with a very different composition and properties is used, though, such as when welding brittle cast irons. These welds are called heterogeneous. The completed welded joint may be termed a weldment.

Classify and define the different types of welding.

- Arc welding: Arc welding is a welding procedure that is utilized to join metal to metal using electricity to make enough heat to melt the metal. Thus, when the liquefied metals cool, it results in a binding of the metals.
- Resistance welding: Resistance welding is a thermo-electric process used to join metals by applying controlled pressure and passing current for a precise length of time through the metal that is to be joined. For example, this welding technology is generally used in the manufacturing industry to join metal sheets and components.
- Gas welding: Gas welding includes using a gas-bolstered flame to heat the metal workpiece and filler material to make a weld. The gas used is commonly a mixture of fuel gas and oxygen, creating a clean, hot flame.
- Solid-state welding: Solid-state welding is a joining process that includes pressure but does not include any liquid or vapor phase. This kind of welding takes place with or without the aid of temperature.
- Thermit welding: Thermit welding is a method of joining heavy sections of steel structures—for example, rails, etc. Since this includes a casting process, using high temperatures as an input, the thermit steel's metallurgical properties make it best suitable for welding steel with high strength and high hardness.

DID YOU KNOW?

What is the role of nitrogen in welding?

Nitrogen is inert, has a low atomic weight, and is inexpensive. As a result, it is used as a purge gas in welding and brazing to avoid porosity in the welding member by preventing oxygen and air from entering the molten metal during the welding process. It is also used for pressure testing and leak detection. Other gases, such as argon, helium, and carbon dioxide, are also used for this purpose, and the gases are released when the flux burns away during stick welding.

- Electronic beam welding: In this welding process, a high-velocity electron beam is passed over two materials that need to be joined. The kinetic energy of the electrons is transformed into heat upon impact, which leads the workpieces to melt and flow together.
- Laser beam welding: Laser beam welding (LBW) is a type of welding procedure used to join metal pieces or thermoplastics with a laser. The beam allows deep welds along with high welding rates. One example of using laser welding is to weld car bodies in automobile industries.

What is a CNC machine?

The acronym CNC stands for Computer Numerical Control. CNC machining is the process of using a computer-driven machine tool to produce a part out of solid material in a different shape. CNC consists of three primary components: a command function, a drive/motion system, and a feedback system.

Metals and Metallurgy

Is metallurgy a part of mechanical engineering or a different domain?

Metallurgy is a domain of materials science and engineering that focuses on metallic elements' physical and chemical properties and their alloys. Thus, it is the art and science of separating metals from their ores and altering them for application. This concerns metals' chemical, electrical, and atomic properties and compositions as well as the rules under which metals are joined to form alloys.

Mining the ore, removing and concentrating the metal or metal-containing compound, and reducing the ore to metal are the three general steps of metallurgy. Other processes are often needed to improve the mechanical properties of the metal or increase its purity.

What is corrosion? How is it caused?

Corrosion is a natural phenomenon that turns a refined metal into a more chemically stable form like oxide, hydroxide, or sulfide. It is the degradation of a material over time due to chemical or electrochemical reactions with its environment.

What is pitting? How is it caused?

Pitting is the kind of nonuniform corrosion that occurs through the entire metal surface but only in small pits. Thus, pitting corrosion is a form of highly localized corrosion that results in small holes in the metal. This is due to the lack of uniformity in metal.

TOOLS AND EQUIPMENT

What is the difference between a machine and an engine?

A machine is a static device utilized for the generation of mechanical power. It has numerous parts, each with a definite capacity (function) that performs a specific task. In contrast, an engine is responsible for running the machine. It is a machine intended to change one form of energy into mechanical energy. Thus, the engine is the core of the machine, which causes the machine to work.

What are gears and types of gears?

A gear is among the most significant power transmission components. It is a rotating machine component that has cut teeth that work with another toothed part, typically having teeth of comparative size and shape, to transmit control. Transmission/gear set refers to the two or more gears working together

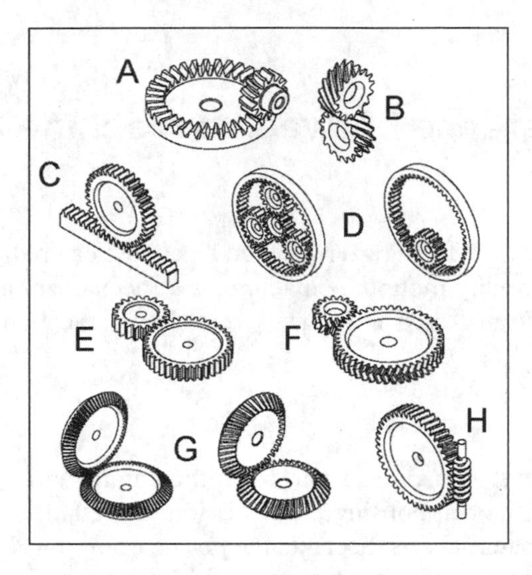

Figure 9: Some common gears include A) bevel gear; B) helical gears; C) rack and pinion gear; D) internal gears; E) spur gears; F) herringbone gear; G) miter gears; H) worm gear.

to deliver "mechanical advantage," which can be utilized to change the speed, torque, direction of rotation, direction of power source, or kind of movement.

Types of gears:

- Spur gears
- Helical gears
- Rack and pinion gears
- Bevel gears
- Miter gears
- Worm gears
- Screw gears
- Internal gears

What are the functions of a gear drive?

The primary functions of a gear drive are to increase torque from the driving motor to the driven equipment, decrease the speed created by the motor, and alter the course of the rotating shafts.

What is a powertrain?

In a motor vehicle, the powertrain (also known as the torque train) consists of the source of propulsion and the drivetrain system, which transfers this energy

DID YOU KNOW?

What is the difference between a machine and a mechanism?

A mechanism is made up of links forming a restricted kinematic chain. Its primary purpose is to transmit or modify motion. A machine is a mechanism concerned with forces that need to be transferred, and it is used to change mechanical work.

into forward movement of the vehicle. Thus, it starts to form the engine and ends at the wheels passing through the bell housing, transmission, driveshaft, and differential. It is a system of mechanical parts that takes the power, or output, of the engine and, through specific gear ratios, slows it and transmits it as torque.

How do mechanisms work? What are different types of mechanisms?

The essential physical or chemical procedures associated with or responsible for an activity, reaction, or other natural phenomena are termed as mechanisms. In engineering, a mechanism is a device that changes input forces and movement into an ideal set of output forces and motion. Mechanisms generally comprise components such as gear trains, chain drives, linkages, brakes, etc. Types of mechanisms include planar, spherical, and spatial mechanisms.

What are turbo machines?

Turbo machines transfer energy between a rotor and a fluid, including both turbines and compressors. In a typical context, turbo machines are systems in which one or more rotating blade rows' dynamic motion contributes energy to or extracts energy from a continually flowing fluid—for example, heavy-duty gas turbines, jet engines, etc.

What are reciprocating compressors?

Reciprocating compressors emit compressed air or gas pulses, which are discharged into the piping that carries the air or gas to its final destination. One

Turbos like this help increase power by forcing air into the combustion chamber.

or more cylinders and pistons that move within them, similar to an internal combustion engine, are the main components of the compressor. These are widely used in the chemical, oil, and gas industries for moving compressible fluids reliably.

What is the difference between a centrifugal pump and a positive displacement pump?

In a centrifugal pump, the flow rate changes with the head, but in a positive displacement pump, the flow rate remains the same.

What are the losses in a centrifugal pump?

The losses in a centrifugal pump are friction losses due to eddies' inflow, leakage, and bearing power losses; disc friction power loss due to friction between the impeller's rotating faces (or discs); and liquid, leakage, and recirculation power loss.

What is the difference between torque and power?

Torque is the rotational equivalent of linear force produced by an engine's crankshaft, i.e., it is the amount of turning force an engine has in one revolution. Mathematically, Torque = Force x Radius.

Power is the speed of an engine that can move a certain weight over a certain distance. Mathematically, Power = Torque x Angular Velocity.

Engines

What is the difference between a heat engine and a heat pump?

A heat engine deals with a cyclic procedure (thermodynamic) and yields a network that converts the heat supply (heat or thermal energy and chemical energy) into valuable work (mechanical).

A heat pump also works on a cycle; however, it consumes work. It is used for maintaining temperatures higher than the surrounding temperature (heating purpose). For example, a heat pump aims to transfer energy to a warm environment, such as a home in the winter.

What is the principle behind the working of combustion engines?

The reciprocating internal combustion engine works on the principle that if we put a small amount of high-energy-density fuel, like gasoline, in a little, enclosed space and ignite it, an incredible amount of energy (expanding gas) is discharged. Thus, in an internal combustion engine, the fuel is being burned inside of the engine itself. Thus, an internal combustion engine is intended to change its pistons' up-and-down movements into the circular motion of the vehicle's wheels.

What are the steps involved in the working of combustion engines?

1. Practically, this process starts when we turn on the vehicle, which utilizes power from the vehicle's battery to get the pistons moving.

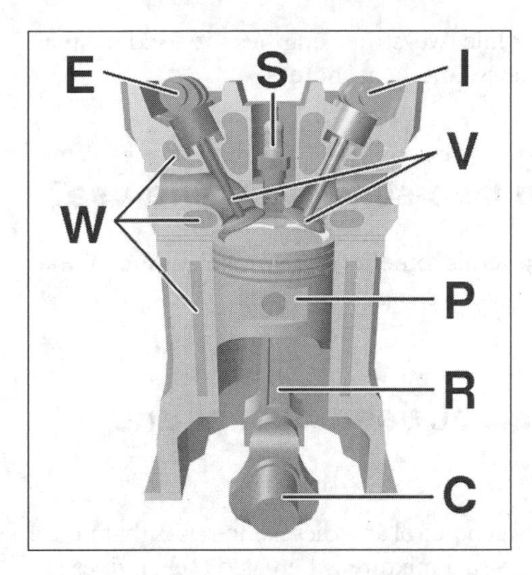

Figure 10: The basic parts of a four-stroke internal combustion engine are the crankshaft (C), exhaust camshaft (E), inlet camshaft (I), piston (P), connecting rod (R), spark plug (S), valves (V), and cooling water jacket (W).

2. When a given piston slides down its chamber, valves open so that fuel and air rush into the vacant space.
3. After the piston goes down to the extent that it can, the connecting rod rotates around the crankshaft and the piston returns up.
4. Since the cylinder valves are closed right now, the air–fuel mixture is crushed into the increasingly smaller space between the cylinder head and the cylinder's highest point. Additionally, the spark plug attachment ignites the compressed fuel and air and the expanding gases push the piston back down into the cylinder.
5. This gives power to the connecting rod to move the crankshaft. Thus, the crankshaft turns, the cylinder valves open, and the piston pushes out the exhaust gases as it rises again in the cylinder. From that point, the cycle continues.

What's the difference between two-stroke and four-stroke engines?

The main difference between two-stroke and four-stroke engines is how quickly the combustion cycle completes. In a four-stroke engine, the piston takes two strokes during each revolution—one compression and one exhaust stroke—followed by a return stroke. On the other hand, a two-stroke engine consists of only one stroke consisting of a compression stroke followed by fuel combustion.

A four-stroke engine is more environmentally friendly with higher efficiency than two-stroke engines and creates a higher torque at a lower RPM. Four-stroke engines are also heavier in weight; they are used in go-karts, internal combustion engines in vehicles, etc., while two-stroke engines are used in small appliances such as lawn tools, chainsaws, and boat motors.

What kind of oil does a two-stroke engine use?

Two-stroke (2T) engine oil is a semisynthetic, low-smoke engine oil that meets the critical requirements of two-stroke engines.

What is the difference between petrol and diesel engines?

The fundamental difference between petrol and diesel engines is that petrol engines use spark plugs to ignite the air–fuel mixture, whereas diesel engines are based heavily on compressed air where the temperature would rise so high, it would cause diesel fuel ignition. Comparatively, diesel engines are more efficient, and diesel fuel is cheaper to purchase. Also, diesel engines are built strong to withstand the high compression of gases; their life cycle is longer than a petrol engine.

What happens if you put diesel into a petrol engine?

Diesel is less volatile than petrol and has a low self-ignition temperature of about 210 degrees Celsius. As a result, the engine will not function properly and will misfire, which will cause the vehicle to generate smoke. After the gas in the fuel lines has been used up, the engine will cut out, and your vehicle may not start again.

What is coolant?

Coolant (also known as cutting fluid) is a lubricant that aids a CNC machine to easily cut materials like metals, fiberglass, and high-density plastics.

How does a crack in a long shaft affect the rotor?

A shaft crack is a rotor fracture that expands slowly. If a fracture develops undetected in an operating machine, the rotor's decreased cross-section would

DID YOU KNOW?

What happens if you put petrol into a diesel engine?

Using petrol in a diesel engine will result in more harm than using diesel in a petrol engine. It increases friction in a diesel engine, meaning that the more fuel that is pumped through an oil burner, the more serious the harm.

be unable to tolerate the dynamic forces imposed on it, which would lead to a malfunctioning of the rotor.

What does air–fuel ratio mean? What does it mean to have a "rich" or a "lean" air–fuel ratio?

A richer mixture (a lower air–fuel ratio) is utilized for acceleration and high-load circumstances to generate cooler combustion products and minimize overheating of the cylinder head, preventing detonation. The ideal air–fuel ratio that burns all fuel without excess air is 14.7:1. This is referred to as the stoichiometric mixture. This ratio is optimal for idle and gentle throttle cruising since it is the most efficient mixture, resulting in the best fuel efficiency and emissions. When we want the engine to start producing more power (for example, when we accelerate), we don't want efficiency; we want power. To produce more power under those conditions, we'll require a 12:1 air–fuel ratio.

When we say "a bit rich," we mean that too much fuel is present, and when we say "a bit lean," we mean that not enough fuel is there. As a result, the air–fuel ratio is wrong and has to be adjusted to restore balance.

COMPUTERS AND PROGRAMMING

Which types of programming language and software are used in mechanical engineering?

- MATLAB: for mathematical programming
- C: for data acquisition and real-time robotic control

DID YOU KNOW?

What is the use of MATLAB for mechanical engineers?

To address issues in control systems, mechanical vibrations, fundamental engineering mechanics, electrical circuits, statics and dynamics, and numerical methods, mechanical engineers use MATLAB. For example, software firms such as Google and Facebook engage mechanical/thermal engineers to maintain effective and safe thermal management of their database and cluster systems. These engineers develop scripts in programming languages like MATLAB/Python and then import them into CFD software to evaluate a variety of ideas.

- C++: an extension of the C language
- Python: robotic tools, which use Python binding
- ANSYS: to design products and semiconductors and create simulations that test product property
- PLC: used to automate industrial equipment
- CAD: for designing
- CAM: for manufacturing processes

What are CAD/CAE software programs and their importance?

CAD (computer-aided design) is used for designing a product. CAD is used by architects, engineers, drafters, etc., to create precision drawings. The most popular CAD software programs are AutoCAD, ArchiCAD, AutoDesk, CATIA, CREO, IronCAD, Keyshot, and Key Creator.

CAE (computer-aided engineering) is used for analysis, testing, and simulating designed visualization. CAE includes finite element analysis (FEA), computational fluid dynamics (CFD), multibody dynamics (MBD), durability, and optimization. Thus, it helps to improve product designs and helps in the resolution of engineering problems for various industries.

What is MATLAB used for in real life?

MATLAB is a programming platform explicitly used by engineers and scientists to analyze and design systems and products that transform our world.

MATLAB allows matrix manipulations, plotting of functions and data, implementing algorithms, creating user interfaces, and interfacing with programs written in other languages. In real life, MATLAB deals with performing numeric computation to analyze and visualize data from different files, programming, and application development. Thus, MATLAB helps to automate many tasks and makes it easy to handle equations of any complexity.

CAREERS

What are the requirements to become a licensed professional engineer in the United States?

Degrees in mechanical engineering usually take four to five years of study. In the United States, most undergraduate mechanical engineering programs are accredited by the Accreditation Board for Engineering and Technology (ABET) to ensure equal standards for all universities. Also, to become a licensed professional engineer, the National Council of Examiners for Engineering and Surveying (NCEES) has set specific requirements where an engineer is required to pass the comprehensive FE (Fundamentals of Engineering) exam and work for a minimum of four years as an Engineering Intern (EI). Also, the individual has to pass the PE (Practicing Engineer) exams.

What are the applications of mechanical engineering?

Mechanical engineering is a broad field that encompasses anything from small, individual parts and machines (such as clocks, microscale sensors, and inkjet printer nozzles) to complex structures (such as spacecraft, ships, and machine tools). Mechanical engineering affects our daily lives directly and indirectly. Examples of this are designing or redesigning all types of machinery, vehicles, and aircraft; developing and testing prototypes of new products and devices; and designing, installing, and managing manufacturing equipment.

What are the different fields (sub-branches) of mechanical engineering?

- Power plant engineering

DID YOU KNOW?

Which materials are used to build a cruise ship?

Aluminum and high-strength steel are used to build a cruise ship. Despite their bulk, lightweight metals keep the center of mass down. The ship's heaviest elements—the engines, propellers, and fuel tanks—are in the lower decks, and despite a ship's top-heavy nature, weight is equally spread to maintain stability. Lightweight products not only tend to cut down on fuel consumption and pollution, but they are also more than capable of performing in the often-harsh climate of the sea.

- Marine engineering
- Aerospace engineering
- Automotive engineering
- HVAC engineering
- Computer-aided engineering
- Mechatronics
- Acoustical engineering
- Manufacturing engineering
- Thermal engineering

What is aerospace engineering? What do aerospace engineers do?

Aerospace engineering is the branch of engineering that deals with the creation, testing, and production of airplanes, rockets, and associated systems and equipment. It has two primary and overlapping branches: aeronautical engineering and astronautical engineering.

An aerospace engineer is responsible for the design, testing, and management of airplanes, spacecraft, satellites, and missiles. They also design propulsion systems, analyze aircraft aerodynamic performance, test prototypes, and ensure that all blueprints, prototypes, and products fulfill technical standards, environmental concerns, client specifications, and operate as designed. Furthermore, they create new technologies for space exploration, aviation, and military systems. (For more information on aerospace engineering jobs and responsibilities see the chapter "Aerospace Engineering")

What is power plant engineering? What are the primary responsibilities of a power plant engineer?

Power plant engineering is the field of power engineering, which focuses on power generation for industries and cities and not for household power development. In a power station, a power plant engineer maintains daily operations. Their primary responsibilities include carrying out operational tests, providing preventive equipment maintenance, testing thermal systems, etc.

What is marine engineering? What are the typical job responsibilities of a marine engineer?

Marine engineering includes the design, development, operation, maintenance, and repair of all major mechanical and engineered equipment aboard a marine vehicle of any kind, such as surface ships and submarines.

The typical job responsibilities of a marine engineer are as follows:

- Preparing comprehensive drawings and schematics as well as system layouts.
- Preparing technical reports, budget plans, contract requirements, and design and construction schedules and presenting them to clients and team members for analysis and feedback. Technical reports include research, systems diagnostics, and design analysis documentation.
- Designing and supervising marine equipment testing, installation, and repair.
- Examining marine gear and equipment and creating work requests and task requirements.
- Conducting testing on marine machinery and equipment for environmental and operational effectiveness, quality assurance, and integrity.
- Establishing relationships with contractors to ensure that the job is completed accurately, on time, and within budget, thus assuring that safety procedures and project deadlines are followed.

What is the role of an HVAC engineer? What are the major job responsibilities of an HVAC engineer?

HVAC (heating, ventilation, and air conditioning) systems are a necessary component of most modern structures. HVAC engineers are the people who develop the systems that regulate air quality and temperature. They are responsible for designing, installing, maintaining, and repairing HVAC systems, which provide thermal comfort and control air quality and temperature. These systems are installed in homes, office buildings, hospitals, stores, schools, etc.

The typical job responsibilities of an HVAC engineer are as follows:

- Approving accessories, including fittings, ducts, water pipes, etc.
- Managing all data center applications as well as installing and maintaining dehumidification systems and different centrifugal chillers.
- Ensuring the availability of fitting tools, including power tools, hand tools, scaffolding, ladder, etc.
- Monitoring site operations, coordinating installation schedules, and providing reports and audits on a daily basis.
- Coordinating with stakeholders to conduct frequent audits and maintaining adherence to all corporate regulations as well as helping in the design of a cost-effective HVAC system that meets all architectural criteria.
- Supporting the commission team with air and hydronic fitting testing, balancing, and adjusting.

An HVAC engineer checks a cooling tower for a large skyscraper in this photograph.

DID YOU KNOW?

What is mechatronics?

Mechatronics is an essential framework in the field of automation and manufacturing. It is an interdisciplinary branch of engineering that creates simpler, smarter, and more reliable systems focusing on mechanics, electronics, and computing. Thus, a mechatronics engineer/specialist deals with robotics, control systems, product engineering, and electromechanical systems.

What is acoustical engineering?

Acoustical engineering is an engineering division concerned with sound and vibration. Typically, acoustical engineers are concerned with regulating noise or distortion that may influence humans' hearing environment. This involves the use of acoustics for the study of sound and vibration.

What do acoustical engineers/specialists do?

Tasks/duties performed by acoustical engineers include:

- Reducing vibrations that influence machinery or equipment.
- Configuring and optimizing theaters and auditorium acoustics.
- Analyzing and improving building acoustics: soundproofing walls, windows, etc.
- Measuring, modeling, and simulating transport-related sound levels.
- Reducing vibration/noise in products such as automobiles, generators, mechanical products, etc.
- Analyzing and reducing noise in industries'/factories' working environments to protect workers' health, avoiding occupational fatigue and other health issues.

What do mechatronics engineers/specialists do?

Tasks/duties performed by mechatronics engineers include:

DID YOU KNOW?

What is metallurgical engineering?

Metallurgical engineers focus on the verticals of various metals' physical and chemical properties and thus deal with all types of metal-related areas. The main branches of metallurgical engineering include physical metallurgy, extractive metallurgy, and mineral processing. The metallurgical engineering course covers mechanical metallurgy, welding metallurgy, hydrometallurgy, etc. In this, students get hands-on experience in gathering and extracting mineral products and learning problem-solving abilities. After the course completion, students are capable of assembling metals to produce materials as per industrial requirements.

- Designing, developing, and maintaining engineering systems.
- Assisting in designing consumer goods; for example, cameras, etc.
- Using CAD software programs to create mechanical design documents for parts, assemblies, or finished goods.
- Maintaining and documenting technical reports.
- Modeling system behavior using software such as MATLAB, Simulink, etc.
- Identifying and selecting design methods for mechatronics systems.
- Assisting with new product development.
- Analyzing effectiveness and efficiency of existing and new mechatronics systems.

What is the meaning of the term solid mechanics?

Solid mechanics is the physical science division that studies the behavior of solid materials, especially the deformation and motion of continuous solid material under applied external loadings, such as forces, displacements, and changes in temperature, that result in inertial force in bodies, thermal changes, chemical interactions, electromagnetic forces, etc.

What is railway engineering?

Railway engineering is a broad engineering field that deals with the planning, designing, and maintaining of all forms of rail transportation systems. It

What is a mechanical reasoning test?

DID YOU KNOW?

Mechanical reasoning is the ability to comprehend mechanical and physical concepts, visual and spatial relations, and good information about tools/apparatus.

Mechanical reasoning tests assess an individual's ability to understand and apply mechanical and physical concepts. These are often used as a pre-employment examination for job placements. Typically, these are jobs such as machine operators, line-assembly workers, electricians, etc. The Bennett Mechanical Comprehension Test, the Wiesen Test of Mechanical Aptitude, and the Ramsay Mechanical Aptitude Test are the most commonly used mechanical comprehension tests.

involves a wide range of engineering disciplines, including civil engineering, computer engineering, electrical engineering, mechanical engineering, industrial engineering, and production engineering. Depending on the role sought within the railway engineering discipline, an engineer could be engaged in the planning, development, repair, or operation of trains and rail systems. This includes monitoring and controlling the rail network and the trains. A rail systems engineer, for example, is responsible for offering expertise and insight into railway projects and systems such as traction power, train, and traffic signal controls; fare collection; rail vehicles; and more.

What kinds of jobs can someone get with a mechatronics degree?

Mechatronics engineers can work in the following areas:

- Robotics engineer/technician
- Automation engineer
- Control system design/troubleshooting engineer
- Electronics design engineer
- Mechanical design engineer
- Data scientist/big data analyst
- Instrumentation engineer
- Software engineer

Other careers for someone with a specialization in mechatronics engineering include:

- Advanced manufacturing and robotics
- Telecommunications and information services
- Agriculture, food, and forestry
- Biotechnology, life sciences, and medical equipment design
- Renewable energy
- Transportation and logistics
- Homeland security and defense

What are the core subjects a student has to study for a degree in mechanical engineering?

- Mechanical Drawing
- Thermodynamics
- Engineering Mechanics
- Solid Mechanics
- Heat Transfer
- Fluid Mechanics
- Machine Design
- Mechanics of Machinery
- Instrumentation and Measurement
- Internal Combustion Engines
- Power Plant Engineering
- Refrigeration and Building Mechanical Systems
- Metallurgy and Materials Science
- Thermal Engineering
- Automobile Engineering
- CAD/CAM
- Mechatronics
- Production Planning and Control

What are the key skills that a mechanical engineer must have?

- Specialized technical skills

- Collaboration
- Critical thinking
- Interpersonal skills
- Commercial awareness
- IT skills
- Teamwork skills
- Ability to work under pressure
- Mathematics
- Creativity
- Analytical skills
- Problem-solving skills
- Relevant technical knowledge
- Verbal and written communication skills
- Good leadership and administration skills

What are some example career pathways for mechanical engineers within the Navy?

The Navy has several distinct focus areas in industrial and mechanical technology, each with its own training pathways and job descriptions. These descriptions are as follows:

- Boatswain's Mate (BM): BMs are responsible for repairing and maintaining equipment in preparation for and during underway operations.
- Electrician's Mate (EM): EMs are responsible for keeping, maintaining, and operating all electrical equipment on ships and also preventing power failures. This includes installing, testing, and repairing advanced electrical equipment and appliances that power the defenses.
- Engineman (EN): The EN's duty includes operating, servicing, and repairing internal combustion engines on ships and small craft. Apart from this, they operate and maintain refrigeration and air conditioning systems, air compressors, desalination plants, and small auxiliary boilers.
- Gas Turbine Systems Technician, Electrical (GSE): GSEs handle massive gas turbine engines' electrical components, which power cruisers and destroyers.
- Gas Turbine Systems Technician, Mechanical (GSM): Mechanical parts of gas turbine engines, main propulsion equipment, and auxiliary propulsion control systems are controlled, repaired, and maintained by GSMs.

- Gunner's Mate (GM): Guided-missile launching systems, gun mounts, and other ordnance equipment, as well as small arms and magazines, are all operated and handled by GMs.
- Hull Maintenance Technician (HT): Metalwork is done by HTs to maintain shipboard structures and surfaces in good working order. They also repair small vessels and maintain shipboard plumbing and underwater sanitation services.
- Machinery Repairman (MR): Replacement parts for ship engines and auxiliary systems are made by MRs using machine tools.
- Machinist's Mate (MM): MMs operate and maintain steam turbines and gears for ship propulsion and auxiliary machinery. They also maintain electrohydraulic steering engines, refrigeration plants, air conditioning systems, and desalination plants.
- Mineman (MN): Underwater mines are detected and neutralized with the aid of MNs. They also test, assemble, and maintain underwater explosive devices and ensure proper repair and performance of mines.

What are the top companies for mechanical engineers to work for?

- National Aeronautics and Space Administration (NASA)
- Google
- The Boeing Company
- Apple
- Lockheed Martin Corporation
- Microsoft
- GE Aviation
- U.S. Department of Energy
- Ford Motor Company
- Walt Disney Company
- Illumina
- Johnson & Johnson
- SpaceX
- Boston Scientific
- John Deere

What are the job responsibilities of a mechanical engineer?

- Assessing project requirements
- Designing or redesigning machines such as electric generators, internal combustion engines, steam and gas turbines, etc.
- Measuring the performance of mechanical components, devices, and motors
- Maintaining and modifying equipment, checking for reliability and efficiency
- Using CAD/CAM software
- Undertaking relevant research
- Providing technical advice
- Analyzing problems and designing solutions
- Analyzing and interpreting data
- Developing and testing prototypes of the devices they design
- Overseeing the manufacturing process for devices

What is the future career outlook of mechanical engineering?

The job opportunities for a mechanical engineer are abundant. Still, one must achieve desired skills and relevant knowledge and training in the latest software tools, particularly for computational design and simulation. In terms of economic growth, it's the only branch of engineering that is immune to the recession's impact.

In the government sector, mechanical engineers can provide their knowledge to different government projects as technical experts. Individuals can likewise work in private engineering companies, providing technical consultancy to both government and corporate firms. Mechanical engineers are responsible for designing power-producing machines such as electric generators, internal combustion engines, steam and gas turbines, and refrigeration and air-conditioning systems. As of May 2019, mechanical engineers' median annual wage was $88,430.

Manufacturing Engineering

What is manufacturing engineering?

Manufacturing engineering is the branch of engineering that deals with the design and operation of systems that create products for everyday use. Manufacturing engineering often combines principles and concepts from other engineering disciplines, such as chemical, industrial, mechanical, and electrical engineering. Consequently, manufacturing engineers possess a unique and specialized skill set that helps them to design, analyze, apply, and operate complex processes and systems.

What are the tasks associated with manufacturing engineers?

Manufacturing engineers understand how the factory floor functions. They create plans and give instructions, including the built details of assembly lines utilizing the tools and machinery on the manufacturing floor. They respond to questions and take care of issues brought by floor personnel. These problems include problems with equipment, work directions, etc. They also work on programming automation equipment/machinery, searching for new approaches to improve procedures, researching and testing new equipment, and conducting ROI analyses. Manufacturing engineers work with product designers to ensure that a product is optimized for manufacturability.

The primary roles associated with their jobs include selecting and designing manufacturing processes, determining the most effective and efficient methods of product fabrication, assembly, and testing. Manufacturing engineers are sometimes also tasked with monitoring procurement, supply chain management and inventories, quality control and management, material flow, and cost estimating and analysis.

What types of problems do manufacturing engineers solve?

Manufacturing engineers tend to work on problems associated with manufacturing processes such as the planning and selection of manufacturing methods, designing and selecting manufacturing equipment, and conducting research to develop the creation of new and improved manufacturing processes. Manufacturing engineers also work in the areas of production planning and control, operations management, production management systems, and materials planning and selection.

What are the subdisciplines of manufacturing engineering?

Manufacturing engineering has several subdisciplines. While these subdisciplines all fall under the broad category of manufacturing engineering, their

Manufacturing engineers research, plan, and design ways to improve manufacturing processes. They might design new tools and machines or find efficient ways to upgrade the processes in creating products.

DID YOU KNOW?

What is kinematics?

Kinematics focuses on the study of the motion of objects and systems without attuning to the force that is causing the motion. In manufacturing, the principles of kinematics are applied when designing mechanisms that are required to move in a particular way. For example, if a robotic arm is needed to move at a certain degree or is needed to provide a desired range of motion, the engineer would use kinematics to analyze how the object or systems move so as to design the best solution to this problem.

focus and application are somewhat different. Additionally, the subdisciplines often merge core concepts from other engineering disciplines. These subdisciplines are mechanics, kinematics, drafting, machine tool and metal fabrication, computer-aided manufacturing, mechatronics, textile engineering, and advanced composite materials. A brief explanation of each subdiscipline is provided below.

What is mechanics?

Mechanics deals with the study and exploration of forces and how their effects are manifested in the world around us. As it relates to manufacturing, mechanics covers the design of a product and how the force exerted on and by the product can affect its structural integrity. Mechanics also helps to determine the most appropriate materials that should be used as well as the manufacturing/machining process that will yield the best results.

What is drafting?

Engineers use various types of drawings for designing and creating products and processes. Drafting is the creation of visual representations of products, processes, or systems. Drawings may be created electronically or by hand. When manufacturing parts for product development, drawings are used to communicate ideas, specifications, required materials, tools necessary, etc.

What are machining and metal fabrication?

Machining and metal fabrication entail the selection and use of machining tools.

What are the four areas of manufacturing engineering?

The Accreditation Board for Engineering and Technology, which is a governing body for all engineering and technology programs, has identified four key areas of manufacturing engineering. These are: 1) Materials and manufacturing processes; 2) Product, tooling, and assembly engineering; 3) Manufacturing systems and operations; and 4) Manufacturing competitiveness.

HISTORY

What is the history of manufacturing engineering?

Manufacturing dates back to before the Industrial Revolution when prehistoric societies created instruments needed for daily living such as mechanisms for fetching water from watering holes and wells, clay pots, clay tablets, and paints as well as tools used for carvings such as the hieroglyphs found in Egypt and other ancient artifacts.

What is the connection between the Industrial Revolution and manufacturing engineering?

It was the Industrial Revolution that spurred the formalization of manufacturing engineering as a discipline. While small-scale manufacturing led to the creation of tools, machines, and everyday artifacts, it was the revolution that introduced more robust manufacturing processes that went beyond traditional methods such as casting and hammering. With the emergence of manufacturing industries came the ability to create more advanced items through the development of the factory system that we still have today.

Who are the pioneers of manufacturing engineering?

Several people pioneered the emerging discipline. Some key figures were Englishman Matthew Boulton (1728–1809) and Scotsman James Watt (1736–1819), who fabricated and assembled a steam engine and later went

One of the most famous pioneers of manufacturing in the United States was Eli Whitney, inventor of the cotton gin, which vastly improved the process of harvesting cotton by separating the fiber from its seeds.

on to commercially produce them. American Eli Whitney (1765–1825) created a cotton gin that provided the opportunity to clean 50 pounds of cotton per day. Englishman Joseph Whitworth (1803–1887) developed measuring machines and standardized the system of uniform measures. It was Whitworth who created the system of uniform screw threads. Frederick W. Taylor is both an industrial and manufacturing engineering pioneer and is known for his invention of the Taylor–White process of creating high-speed cutting tools. Taylor's work provided groundbreaking knowledge about the many aspects of cutting metal, which minimized error and led to more robust machining processes.

TOOLS

What tools do manufacturing engineers use?

- Pulsed fiber laser welding system: This is a high-powered, high-speed laser used for welding plastics and metals.
- Drill press: A drill press is also known as a drilling machine. The press removes materials from a stationary workpiece by making vertical holes

with a rotating tool. Drill presses come in many different types, including hand-feed machines, portable drill presses, upright drill presses, and sensitive drilling machines (small machines used for drilling small holes). The drill presses vary in size, which is determined by the drill needed, the size of the workpiece, the amount of power available, the floor space required to complete the task, and other factors.

- Mill: A milling machine is used to remove material from a workpiece that is moving. The cutting tool of the mill rotates as the workpiece is moved along an axis.
- Lathe: A lathe is a large machine that rotates a workpiece to precisely machine it. The workpiece is typically made of hard materials such as metals and plastics. It allows for the workpiece to be rotated on an axis so that it can be manipulated. The lathe allows the workpiece to be cut, sanded, knurled, drilled, and deformed symmetrically (in the same way on all sides) because it rotates. Different types of lathes include speed lathes, engine lathes, bench lathes (used for small precision workpieces), toolroom lathes, capstan lathes, and turret lathes (mass-production workpieces).

What are common drill press operations?

Simply put, the purpose of drilling is to use a drill bit to cut a circular hole through a workpiece. Several operations can be performed using a drill press. The most common are drilling, reaming, lapping, boring, counterboring, countersinking, spot facing, and tapping.

A drill press is simply any machine process that drills holes into materials.

Computers

What is 3D modeling?

Three-dimensional modeling is a computer graphics technique. Objects and parts are created in a computer-aided design (CAD) software. The designs are used for manufacturing parts, video games, film special effects, construction, architecture, and the medical industry.

What is computer-aided design?

Computer-aided design (CAD) is the utilization of a wide range of computers or workstations that help engineers, architects, and other design professionals in the creation, modification, analysis, or optimization of their design activities. Thus, CAD software is utilized to increase designer productivity, improve design quality, improve communications through documentation, and make a database for manufacturing. CAD is broadly utilized in numerous applications, including car, shipbuilding, and aviation businesses; modern engineering structures; prosthetics; and more.

What software programs do manufacturing engineers use for CAD?

- Inventor
- Fusion 360

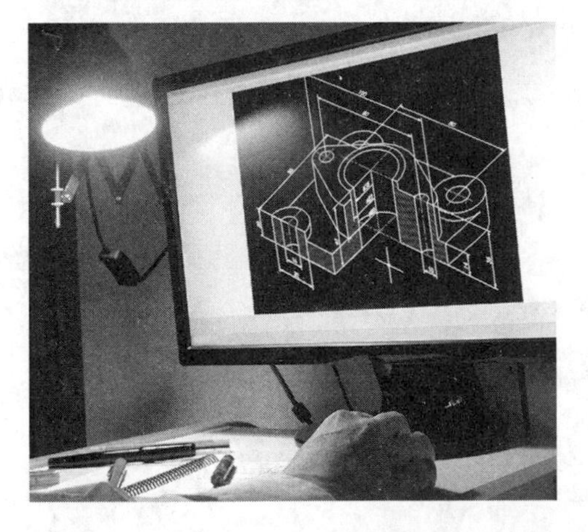

Thanks to computers, engineers have an amazing way to design parts using computer-aided design (CAD).

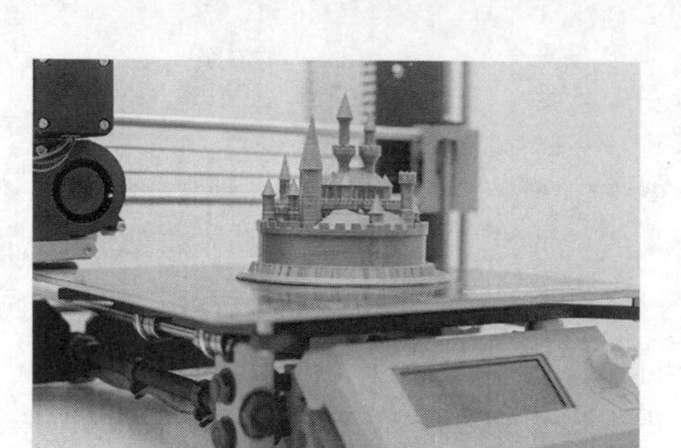

3D printers can make just about anything you might imagine, from simple tools to models, prototypes, and even houses.

- AutoCAD
- CATIA
- Sketchup
- TinkerCAD

What is 3D printing?

3D printing is also called additive manufacturing; it essentially prints three-dimensional, solid models from a 3D model using a printer that uses filaments instead of ink. The filament is the material that is used to print the 3D model. Prints are completed layer by layer, which is also termed layered manufacturing, and this is the process of turning digital designs (CAD models) into three-dimensional objects by depositing successive layers of material. Materials such as metal, ceramic, and polymer are used in this process to manufacture parts that often cannot be produced by any other manufacturing technology. One of the technologies used by today's 3D printers is called selective laser sintering (SLS). During SLS, tiny particles of plastic, metal, ceramic, or glass are fused by heat from a high-powered laser to form a solid, three-dimensional object layer by layer.

MANUFACTURING

What is advanced manufacturing?

Advanced manufacturing is the integration of new innovative technology and techniques to improve both product design and production processes with the relevant advanced technology to facilitate cost-effective and competitive products. These production processes highly depend on automation, networking, computation, and information. Thus, advanced manufacturing integrates the most up-to-date machinery with processes to add value and create highly differentiated products.

What is agile manufacturing?

Agile manufacturing is a strategy that focuses on responding quickly to the needs of the customer. The goal is to provide personalized products at unprecedented speeds while controlling the overall costs and maintaining high-quality standards. Industries that use agile manufacturing develop platforms for the designers, marketers, and production workforce to share information and updates about parts and products, production capacities, and problems, particularly concerning the quality of a product, to ensure that customers' needs are met.

What are the key elements of agile manufacturing?

The key elements of agile manufacturing include:

- Modular product design: Products are broken up into small subassemblies (modules) that can be built independently and used in diverse ways.
- Information technology: With advancements in manufacturing processes and increased automation and technology, more and more information (data) is generated about the needs of the customers, the current state of the markets, and the materials/products. In order for industries to be agile, they must quickly analyze the information and disseminate it around the company in order to ensure that they have a fast response to changes in the customer and market needs.
- Corporate partners: Agile industries must create partnerships with other organizations and industries that will help them acquire materials and resources quickly in order to meet changing customer needs.
- Knowledge culture: an agile industry or organization must create an internal culture with their employees that resembles flexibility and adaptability. These organizations typically invest in ongoing employee training.

What is automation?

Automation is the technology that helps to perform a procedure with minimal or no human assistance. People often think about robots when the concept

DID YOU KNOW?

Why would an industry change from traditional manufacturing to agile manufacturing?

Rapidly evolving environments, constant technological developments, increased access to information, and workforce transformations all led to major shifts in the manufacturing industry and the adoption of agile manufacturing. Nowadays, companies are working in a highly competitive environment where a small decline in performance, product quality, or product delivery can have a huge effect on a company's survival and reputation among consumers. Through agile manufacturing, companies take a competitive advantage while focusing on rapid responses to customers and making fast changes based on customer demand.

of automation comes up in conversation. It is used in various control systems for operating applications ranging from a simple on–off control, such as a household thermostat controlling a boiler, to a multivariable, high-level algorithm, such as a large industrial control system. These are the four different types of automation:

- Basic automation: This is typically the use of software to manage business practices. For example, companies may use software that creates, sends, and processes invoices and sends them to the right person at the right time. The tasks that the software completes are usually repetitive and would be done on a daily basis.

- Process automation: A robotic process automation system can log into software or applications, move files, fill in forms, perform basic internet searches, make calculations, and even open emails. This will typically reduce the number of employees needed to complete some tasks and can lead companies to save money. Ideally, this would enable some employees to be moved to higher-level positions.

- Integration automation: At this level, humans define the machine function, the boundaries within which it will operate, and the tasks it will complete. The machine will be able to mimic human tasks. Through intelligent automation, the industry can increase the speed and scale of work completion and can do more work than if it were just humans performing the work. The company/organization would examine their process and identify the ways to best use the human worker's support and high levels of thinking. If the task can be done without a human, then it will; otherwise, it will just include the human when necessary.

- Artificial intelligence (AI) automation: This is the most complex level of automation. At this level, machines can learn and make decisions. Basically, they keep track of all the previous situations, analyze the actions and results, and then apply the old information to the new situation. An example of this is the virtual assistant/virtual chat feature that companies employ as customer service agents on their websites.

What are the manufacturing applications of automation and robotics?

Manufacturing and production are the most important applications for automation. Automated production systems are classified as either fixed automation, programmable automation, or flexible automation.

- Fixed automation: Fixed automation works best in a production environment that is sequenced and where comments can be preprogrammed into machines. It is not easily changed from one product to another because the programming involves intricate aspects of the machinery like gears, wiring, and electronics. This can be seen in large-volume manufacturing industries such as the automotive industry.
- Programmable automation: Programmable automation is best suited for batch production because for each batch, the machines are reprogrammed. This means that periods of nonproductivity occur. An example of programmable automation is industrial robots that can be reprogrammed to perform different tasks in between regular tasks.
- Flexible automation: Flexible automation is an extension of programmable automation. A limited product variety allows for reprogramming to occur more quickly and even off the production line. This allows for

A robotic arm mimicking the motions of a human to assemble something would be an example of integration automation. Such robots only repeat a set of motions that have been programmed into them. The next step beyond that is AI automation, in which a machine can actually learn and improve a process.

other technologies to be put to use during the "downtime" of the machines being reprogrammed. This allows for a mixture of products and processes to be used on the same production lines.

In the industrial workplace, automation helps in improving productivity and quality, reducing errors and waste, and making the manufacturing process flexible, thus increasing safety, reliability, and profitability. Automation can be achieved by various means, including mechanical, hydraulic, pneumatic, electrical, electronic devices and computers, etc. The benefits of automation include labor savings with lower operating costs; faster return on investment (ROI); increased production output; improvements to quality, accuracy, and precision; and reduced factory lead time.

What is benchmarking?

Benchmarking is a process in which organizations continually seek to improve their practices. In this process, the performance of a company's products, services, or procedures is measured against those of another business that is viewed as the best in the business, otherwise known as "top tier."

What is the purpose of benchmarking in engineering?

The purpose of benchmarking is to recognize internal opportunities for development. By considering organizations with superior performance, separating what makes such prevalent performance possible, and then comparing those procedures with how your business works, you can implement changes that will yield significant improvements. Benchmarking can enable companies to gain an independent view about how well they perform contrasted with different organizations, help to recognize zones for continuous improvement, develop a standardized set of procedures, monitor organization performance, and set performance expectations. It is typically performed because of requirements that emerge inside an organization.

How does an engineer benchmark?

Benchmarking is a straightforward but detailed process that involves the following steps:

- Choose an item, service, or internal department to benchmark.

Why would a manufacturing engineer benchmark?

DID YOU KNOW?

Benchmarking is the practice of contrasting business procedures and performance metrics with industry bests and best practices from different organizations using a specific indicator, bringing about a metric of performance that is then contrasted with others. Thus, benchmarking enables you to concentrate on best practices from your rivals. It enables you to get point-by-point comparisons between organizations. A manufacturing engineer can use this information to make improvements on the production line (such as which type of automation is appropriate) that could save time and money.

- Determine which top-tier organizations you should benchmark against: which organizations you'll compare your business with.
- Gather data on their internal performances.
- Compare the information from the two associations in order to recognize holes in your organization's performance.
- Adopt the procedures and approaches set up inside the top-tier performers.

What is a bottleneck?

The term "bottleneck" typically refers to the neck or mouth of a bottle and the fact that it is the narrowest point in a bottle and the most likely place for a blockage to occur, slowing down the flow of liquid from the bottle. Manufacturing engineers define a bottleneck as a point of congestion/blockage that emerges when workloads arrive at a given machine/operation more quickly than that machine/operation can handle them. Or, more plainly, it is the step in the manufacturing process that takes the longest time to complete. This step, like the bottleneck in a bottle, slows down the flow of materials.

How does a bottleneck impact manufacturing?

A bottleneck significantly impacts the flow of manufacturing; it can create major delays and increase the expense of production. Organizations are more in danger of bottlenecks when they start the production procedure for new items in light of the fact that the process might have imperfections that must be recognized and adjusted; this circumstance requires more investigation and tweaking. Bottlenecks may likewise emerge when demand spikes unexpectedly and surpasses the production limit of a company's industrial facilities or suppliers.

What is an example of a bottleneck?

Recently, Tesla confronted a bottleneck with its Model 3 preorders creating a backlog in production time. In 2017, Tesla's CEO set a goal of producing 20,000 Tesla Model 3 vehicles each month, but 500,000 people reserved the vehicle. When the company was unable to produce the vehicles at this rate, it was discovered that the bottleneck in production was the battery quality and Tesla's ability to produce enough batteries without defects.

Do different types of bottlenecks exist?

Bottlenecks in manufacturing can be either short-term or long-term. Short-term bottlenecks are temporary and are not typically a significant issue. A case of a short-term bottleneck would be a skilled worker taking a couple of days off. Long-term bottlenecks happen constantly and can significantly hinder production. A case of a long-term bottleneck is the point at which a machine isn't effective enough and thus has a long line. Thus, identifying bottlenecks is basic for improving the manufacturing production process since it enables the engineer to decide and correct where accumulation happens.

Describe traditional manufacturing environments.

In traditional manufacturing environments, similar machines are kept near one another (for example, lathes, mills, drills, etc.). These formats are pro-

Tesla had a problem fulfilling orders for the Model 3 (pictured) when it had an issue with battery defects.

Describe subtractive manufacturing.

Subtractive manufacturing is the construction of 3D objects by cutting away from a concrete block of material. It involves cutting/hollowing or extracting parts from a block of material usually made of metal, plastic, or wood. It is performed manually or, more often, powered by CNC machines.

DID YOU KNOW?

gressively vital to machine breakdowns, have standard jigs and fixtures in the same areas, and support significant division levels.

Describe cellular manufacturing.

Cellular manufacturing systems are divisions of just-in-time manufacturing and lean manufacturing. Here, production workstations and equipment are arranged in a sequence by the groups of parts created. This significantly improves the material flow by reducing the distance traveled by materials with minimal transport or delay and increases production velocity and flexibility while reducing capital requirements.

Advantages of cellular manufacturing include reducing setup time, less material handling cost and time, improvement in machine utilization and product quality, less wastage, and simplifying process routing and supervision.

Describe computer-aided manufacturing.

Computer-aided manufacturing (CAM) is defined as the utilization of software and computer-controlled machinery to automate (plan, manage, and control) a manufacturing plant's process. CAM refers to an automation procedure that precisely converts the product design and drawing of an object into a code format readable by the machine to fabricate the item. Some of the CAM software programs available are CATIA, CAMWorks, Fusion 360, hyperMill, Mastercam, and Powermill.

Total Quality Management

What is total quality management (TQM)?

Total quality management (TQM) depicts an organization's approach to long-term success through customer satisfaction. It works on the principle of a participative and systematic approach based on organization-wide, continuous planning; implementing organizational improvements; and detecting and eliminating errors in the manufacturing process.

In a TQM effort, all individuals from an association participate in improving processes, products, services, and the environment wherein they work. Thus, the primary focus of TQM is identifying problems and improving customer experience and satisfaction by consistently meeting customer expectations.

Define some of the key elements of TQM.

- Customer-focused: A consumer decides quality standards. Regardless of what a company does to encourage quality enhancement, the client decides whether the initiatives are worthwhile. Companies' initiatives include recruiting personnel, incorporating quality into the design process, and upgrading hardware or technology.
- Total employee involvement: All workers engage in contributing to shared objectives. Complete employee participation is achieved only when fear is removed from the workplace and empowerment has created the right environment. High-performance work systems leverage continuous efforts to enhance regular company operations. Thus, one type of empowerment is self-managed work teams.

In total quality management, the company involves all employees in improving processes in order to guarantee customer satisfaction.

- Process-centered: Process thinking is a central component of TQM. It is a sequence of steps that converts suppliers' resources (internal or external) into outputs delivered to customers (internal or external). The actions appropriate for carrying out the procedure are established to monitor unforeseen changes, and performance measurements are observed continuously.
- Integrated systems: Every organization has a distinct work environment. Unless a good quality culture has been cultivated, it is practically impossible to succeed in its goods and services. Thus, to continuously enhance and meet the needs of consumers, personnel, and other stakeholders, an integrated system is of utmost importance. Therefore, microprocesses build up more extensive processes, which further aggregate into the business processes needed for plan development and execution.
- Strategic and systematic approaches: Strategic planning, or strategic management, involves developing a strategic approach that integrates quality as a core element. This approach helps in meeting the vision, purpose, and goals of an organization.
- Continual improvement: Continual improvement encourages a company to be both analytical and innovative in identifying opportunities to become more sustainable and more successful in achieving stakeholder needs.
- Fact-based decision making: To enhance decision-making effectiveness, achieve consensus, and facilitate prediction based on historical data, TQM allows an organization to collect and analyze data consistently.
- Communications: Good communication plays a significant role in sustaining productivity and inspiring employees at all levels during periods of organizational transition and as a part of day-to-day activities. Strategies, processes, and timeliness are essential parts of communication.

What is lean manufacturing?

The term "lean manufacturing" was derived from Toyota's 1930 operating model *The Toyota Way* (Toyota Production System [TPS]). Lean manufacturing intends to increase the value of products delivered to the customer while focusing on cutting out waste, ensuring quality while solving the customer's problems. It makes the business more efficient and responsive to market needs.

The five critical lean principles are:

1. Value: Concisely specifying value by a particular product
2. Value stream: Describing the value stream for each product
3. Flow: Making value flow without interruptions

DID YOU KNOW?

What is quality assurance?

Quality assurance (QA) is a way of maintaining a desired level of quality in a service or product and reducing defects in manufactured products. This is done especially by giving attention to every stage of delivery or production. Quality assurance provides the customer confidence that quality requirements will be fulfilled.

4. Pull: Allowing the customer to draw value from the producer
5. Perfection: Seeking perfection

What is Six Sigma?

Six Sigma is a data-driven approach to improve the quality of the manufacturing process and thereby the product as a whole. Six Sigma strategies seek to ensure near perfection of the manufactured product and has its roots in statistically explaining how a manufacturing process is performing. The measure of Six Sigma is determined by a process that does not produce more than 3.4 defects per million. The two Six Sigma methodologies are called DMAIC and DMADV. DMAIC is used when the goal is to improve an existing process, while DMADV is used when the goal is to create new products or processes.

What is DMAIC?

Define, measure, analyze, improve, and control (DMAIC) is a data-driven quality strategy used to improve processes.

1. Define the problem, opportunity for improvement, customer's voice and their requirements, project goals, and customer (internal and external) requirements.
2. Measure process performance.
3. Analyze the process to determine root causes of variation and poor performance.
4. Improve process performance by addressing and eliminating the root causes.
5. Control the improved process and future process performance.

DID YOU KNOW?

What is the difference between lean and Six Sigma manufacturing?

Both lean and Six Sigma manufacturing work toward eliminating waste and creating efficient processes but take different approaches to accomplish it. Lean manufacturing focuses on analyzing workflow in order to decrease the processing time and eliminate waste; it aims to optimize value to the customer by utilizing as few resources as possible. Six Sigma manufacturing aims for perfect outcomes, which can minimize costs and obtain desired customer satisfaction levels.

Lean and Six Sigma manufacturing both work toward enhancing procedures, increasing product quality, and improved customer experience. The critical distinction is that lean manufacturing looks for opportunities to improve flow, whereas Six Sigma manufacturing relies on producing good outcomes.

What is DMADV?

DMADV stands for:

1. Define the design goals.
2. Measure and identify characteristics that are critical to quality.
3. Analyze the data to identify alternative design approaches.
4. Design the alternative.
5. Verify the applicability of the design.

What is Design for Six Sigma (DFSS)?

1. Define design goals that are consistent with customer demands and the enterprise strategy.
2. Measure and identify CTQs (characteristics that are Critical to Quality), product capabilities, production process capability, and measure risks.
3. Analyze to develop and design alternatives.
4. Design an improved alternative that is better suited as per the analysis in the previous step.

DID YOU KNOW?

What is the Kaizen process?

The word *kaizen* means "continuous improvement" in Japanese. It refers to business operations that consistently enhance all functions and include all of the workforce from the CEO to the assembly-line workers. The minor improvements used in the Kaizen process may involve quality management, just-in-time distribution, standardized work, reliable machinery, and the elimination of waste. The core concept is that improvements can come from any employee and that everybody has a role in the company's future. Everyone should try to help make the company better. Thus, it improves quality, productivity, safety, and workplace culture.

5. Verify the design, set up pilot runs, implement the production process, and hand it over to the process owner(s).

What is just-in-time manufacturing?

Just-in-time (JIT) manufacturing, also known as the Toyota Production System (TPS), was first developed by Taiichi Ohno to meet clients' demands with minimum delays. It is a workflow methodology aimed at reducing setup times, manufacturing cycle times, and lessening response times from suppliers and clients. This helps organizations control their processes' variability and increase productivity while keeping fewer costs. Some examples of companies successfully implementing JIT are Toyota, Dell, McDonald's, and Harley Davidson.

What is supply chain management?

Supply chain management combines organizational units along a supply chain and coordinating information, financial flows, and material to fulfill customer demand. A supply chain includes manufacturers, suppliers, transporters, warehouses, retailers, and customers. Successful supply chains manage flows of products, information, and funds to provide a high level of product availability to the customer while keeping costs low.

The main supply chain stages are customers, retailers, wholesalers, distributors, manufacturers, and suppliers. The main objective of a supply chain is to maximize the net value generated.

What is supply chain profitability?

Supply chain profitability is the total profit to be shared across all stages of the supply chain. Success should be measured by the total supply chain surplus, not profits at an individual level. Supply chain decisions significantly impact each organization's success or failure because they significantly influence both the revenue generated and the cost incurred.

Supply Chain Surplus = Customer Value – Supply Chain Cost

What are robotic material handling and tending systems?

Robots perform various programmed tasks in assembly and production settings. They are regularly used to perform work that is hazardous or unacceptable for human laborers such as repetitious work that causes boredom and could lead to injuries due to the worker's inattentiveness. In the manufacturing industry, material handling refers to the movement of production components by robotic arms, usually on or off a conveyor belt or to group together for production. Machine tending is similar but more precise, referring to a robotic arm that loads and unloads a stationary production machine.

Robots can work around the clock with limited operational downtime intervals for maintenance and high consistency levels, whereas manual material handling and tending are slow, inconsistent, and less efficient. Thus, material handling with robots can automate the most repetitive, mind-numbing, and unsafe tasks in a production line. For example, robot duties include part selection, moving parts, packing, palletizing, stacking, emptying, and machine feeding.

What is process manufacturing?

Process manufacturing is a branch of manufacturing that deals with formulas and manufacturing recipes. It manufactures goods by combining supplies and raw materials, utilizing an equation or formula. For example, it is mainly used in industries that manufacture bulk amounts of products such as food, refreshments, refined oil, gas, pharmaceuticals, synthetics, and plastics.

Thermal or chemical conversions, such as heat, time, or pressure, are often needed in the production process. As a result, it is impossible to disassemble a substance produced by process manufacturing into its constituent parts. For example, a soft drink cannot be further broken down into its components once it is made.

PRODUCT DESIGN

What is product design?

Product design is a creative and systematic approach to ensure that an object's specifications and characteristics meet its desired need. This process takes into consideration the criteria and constraints that govern the usability of the product. The design process captures every decision made from the initial product conception to the final product's final creation.

What is product quality?

Product quality incorporates features that can satisfy customer needs by improving the performance quality of a product in conformance to specifications where conformance means quality consistency and freedom from any defects. Product quality has two dimensions: level and consistency. From a manufacturing point of view, product quality is achieved with quality control and quality assurance.

What is product life expectancy?

Product life expectancy is defined as the life cycle of a product. It is calculated as the length of time from when a product is launched into the market

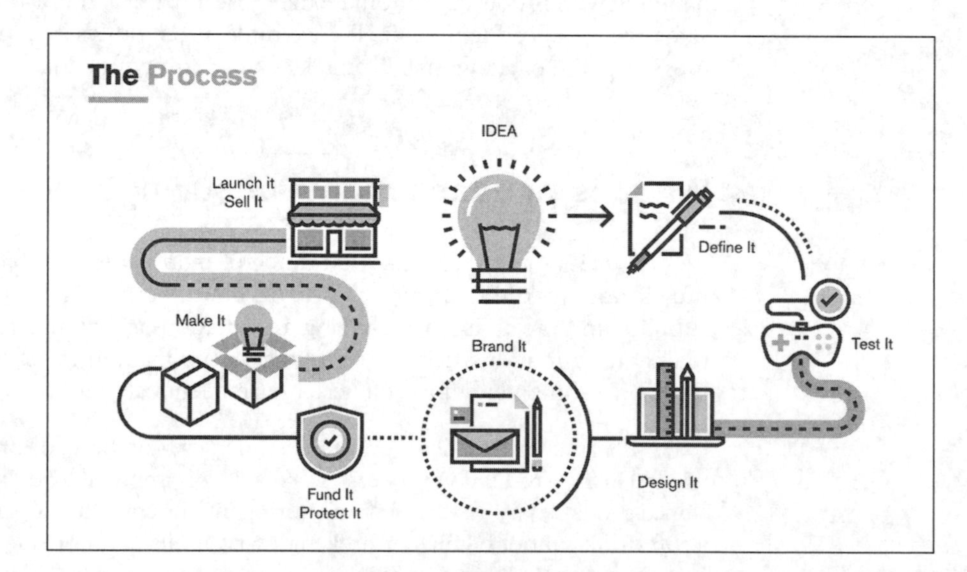

Figure 11: There is a definite process in manufacturing engineering that takes an idea from original concept to final, sellable product.

until it's removed from the shelves. A product's life cycle has four stages: introduction, growth, maturity, and decline. Life expectancy is based on industry, product, and market factors and varies accordingly. Also, recent developments in technology and society have some influence on a product's life span.

What is the role of technology in product design?

Technology's role in product design is very important. It integrates art, science, and technology to create new working products. Enhanced technology has also made it easier to implement, remanufacture, and design the products. Technology influences the way people engage, read, and learn. Computer modeling and simulations, for instance, have helped several businesses and designers to test design features before investing in development tools. Technology has, therefore, helped turn exciting ideas into actual goods and refine them.

What is prototyping?

In manufacturing, it is common to create replicas of a product to determine if it meets the needs and specifications for which it was created. Creating a product can be very expensive and time intensive. Sometimes, it is necessary to develop cheaper but similar product models that can withstand testing to determine if possible modifications are required.

Prototyping is the process of making a model of an actual product. Prototyping may include the use of actual tools and materials from which the product will be made or computer-aided techniques, such as 3D printing with polymers, to create the scaled model.

Prototyping can be physical or virtual. Virtual prototyping involves using software models, such as virtual reality, and advanced graphics that allow designers to assess the product in detail and from different views.

What does DFMA mean?

DFMA (Design for Manufacture and Assembly) is the combination of two systems: Design for Manufacture, which is designed to simplify the production of the parts that will form a product, and Design for Assembly, which is meant to develop the product's simplicity of assembly. Thus, DFMA helps the design team simplify the product's structure, eliminate waste, and diminish the manufacturing and assembly expenses.

DID YOU KNOW?

What is green design and manufacturing?

In the manufacturing field, green design and manufacturing, or greening of manufacturing, is the reestablishment of production processes and operations with an environmentally friendly approach. In this environment, workers utilize fewer natural resources and reduce pollution by using recycled and reused materials in their processes. Thus, green manufacturers explore, create, and use technologies to reduce the impact on the environment.

What is meant by service design?

Service design is an activity mainly concerned with the planning and organizing of people, infrastructure, communication, and material segments of service to improve its quality and the interaction between the service provider and its clients. These improvements are based on both the necessities of clients and the skills and capabilities of service providers.

How are materials selected in the manufacturing process?

In manufacturing, materials selection is an essential step in order to accomplish the machine's reliable service and functionality. It is based on the material properties that can optimize product performance, reliability, and cost. Thus, the selection process includes identifying the design requirements and materials selection criteria and then identifying and evaluating the candidates' materials. Design requirements are comprised of performance, reliability, cost, government regulations, and industry standards. Once these requirements are fulfilled, then the selection of materials takes place.

What are some material properties?

Material properties include flexibility, strength, weight, durability, resistance to heat and corrosion, ability to be cast and welded, electrical conductivity, and machinability.

What are the different types of manufacturing processes?

The four main types of manufacturing are molding, machining, joining, and shearing/forming.

Molding involves the following processes:

1. Casting: Involves heating plastic, then pouring it into a mold until it becomes liquid. The mold is withdrawn once the plastic cools, creating the desired form.
2. Injection molding: Melts plastic to produce 3D products such as butter tubs and toys.
3. Blow molding: Used for the manufacture of piping and milk bottles.
4. Compression molding: Used for large-scale items such as automotive tires.
5. Rotational molding: Can be used for furniture and shipping drums.

Machining involves processes like turning, drilling, milling, and grinding. Some examples are the use of tools such as saws, shears, and rotating wheels to obtain the desired result; laser machines' use of a high-energy laser beam to cut a piece of metal; and plasma torches transforming gas into plasma using electricity.

Joining uses processes like welding, soldering, mechanical fasteners, epoxy, etc., to combine materials.

A worker operates a plastic injection molding press in this photo. Many plastic parts you see in your daily life are created using this process.

Shearing allows straight cuts through a piece of metal using cutting knives.

Forming is another metal-shaping method, which uses compression or another kind of stress to move materials into the desired shape.

How are manufacturing processes selected?

The choice of which manufacturing process to select is based on many factors. These factors are termed as process selection drivers. These include the quantity of a product, equipment cost, manufacturing cost, maintenance cost, processing time, level of skilled labor required, energy consumption and source, material availability and price, waste produced, product size/quantity, and design tolerance.

What are the main elements of computer-integrated manufacturing?

- Computer numerical control
- Adaptive control
- Industrial robots
- Automated material handling
- Automated assembly systems
- Computer-aided process planning
- Group technology
- Just-in-time production
- Cellular manufacturing
- Flexible manufacturing systems
- Expert systems
- Artificial intelligence
- Artificial neural networks

What are the different types of costs in a manufacturing environment?

Cost	Description
Fixed Costs	These are costs that do not vary with output.
Sunk Costs	These are costs that cannot be recovered on leaving the industry, e.g., advertising.

Cost	Description
Variable costs	These are the costs of how much is produced (e.g., raw materials). Variable costs vary with output, i.e., as output increases, more variable costs will occur. For example, more cars' production will lead to more costs of raw materials, such as metal, tires, and plastic.
Semivariable Costs	Some costs may contain both fixed and variable factors. A business, for example, may continue to hire employees, even during a production slump. But they can still take on more employees or pay overtime as production increases—for example, costs like labor, which to some extent depend on output.
Marginal Costs	These are the costs of producing an extra unit. They consist of short-run costs and long-run costs.

What are short- and long-term costs?

In the short term, a firm will have fixed capital but can vary the quantity of labor. However, an organization is expected to see declining marginal returns in the short term. This suggests that if an organization recruits more employees, they will have a stage where a diminishing marginal product will occur for new workers. In the long term, all aspects of production, such as capital and labor, may be varied by a company. Therefore, the company will not face declining returns.

What factors affect the total cost of manufacturing a product?

- Raw materials: The final price of a product will be influenced by the kind of raw materials selected. Also, the customer can choose from thousands of plastic resins and metal alloys. They all have specific physical properties and come at varying price points. To make the product more robust, it is always best to spend a bit more on the raw materials' quality.

DID YOU KNOW?

What is meant by outsourcing?

Outsourcing is a cost-cutting business practice. One company hires another company to perform services and create goods traditionally performed internally by the company's employees and staff. This sometimes involves transferring employees and assets from one firm to another. Some standard outsourcing practices include human resource management, research, supply chain management, accounting, marketing, customer support, and service.

- Tooling: The cost of the tool used to make a product is part of the price. The cost of a tool is determined using several parameters. The parameters include the grade of steel needed, the number of shots it should last, etc.

- Part complexity: This refers to the complexity of a particular part and the number of production phases and distinct procedures needed to achieve the final design. Cost increases with each additional process. This happens due to increasing the manual labor involved in the setup, testing, measuring, and even maintaining tight tolerances on multiple features. Thus, wherever feasible, it is best to keep designs simple in order to avoid the extra cost.

- Volume: The cost of manufacturing tools is the same regardless of whether they are used to produce one or even 1 million parts, although with volume, the cost per finished component varies inversely. The price for each part gradually decreases as more pieces are made from a particular tool, so this is one approach to amortize the initial cost.

- Precision: Accuracy is the proximity of a feature's measurement to the given value that one is striving to achieve. At the same time, precision means being accurate over and over again. It is the closeness of the measurements to one another. Temperature, humidity, and other environmental factors influence the raw materials. Thus, we need to monitor both the workpiece and the environment to significant degrees to be precise.

- Labor: Trained operators and technicians who are highly skilled and bring value to the finished product are necessary to make complex components. People are required to manufacture and inspect products; run machinery; and pack, load, and deliver the products.

What is cost estimation?

Cost estimation in project management predicts the cost and other resources necessary within a given framework to complete a project. For any element needed for the project, cost estimation determines a total amount that defines a project's budget. Thus, a cost estimate is a prediction of how much money is needed to successfully complete a project.

What are the different types of cost estimation?

The three types of cost estimation, as per their scope and accuracy, are order of magnitude estimate, budget estimate, and definitive estimate.

1. Order of magnitude estimate: This is a rough approximation of costs, particularly during the evaluation and preparation phase, that is used at the very early stage of a project. The objective of this method of cost estimation is to provide an understanding of general and overall expenditures. It is done using historical data from a similar project or an aggregation of data from multiple projects. Also, in some cases, no data is used at all.
2. Budget estimate: This is usually done during the project's actual planning stage. It has more accuracy (–10 to +25 percent) than an order of magnitude estimation. This method produces a preliminary list of costs based on the critical components of the project. This is, however, achieved during the project's actual planning stage.
3. Definitive estimate: This is also regarded as a detailed cost estimate. It has the highest accuracy (–5 percent to +15 percent). Reports produced from a definitive estimate include an exhaustive list of project deliverables with brief details for clarification and understanding.

What are the principles of cost estimation?

1. Integrity: Any cost estimate should be generated with a high ethical integrity standard and by following an effective and transparent process.
2. Information accuracy and relevance: The cost estimate should be developed based on the best information available, i.e., engineering judgment and professional guidance should be taken before assuming the estimate.

3. Uncertainty and risk: Project cost should be identified and listed with an estimate with consideration to potential risks or uncertainties affecting the project.
4. Expert team: This principle assumes that only a skilled, professional interdisciplinary team should generate cost predictions. To build the budget estimates, the experienced team needs to use statistical methods and tools.
5. Validation: Experienced professionals such as project managers should develop initial estimates and submit them to an unbiased team in order to validate cost predictions. This second independent judgment helps make the predictions more reliable and capture multiple perspectives on the estimating process.

What is work simplification?

Work simplification aims to establish better working strategies that optimize performance while minimizing investment and costs. Work simplification provides an organized approach to improve operations. This makes it possible to perform everyday activities while reducing the number of resources needed to accomplish a job.

What are the steps involved in work simplification?

Typically, work simplification uses the following six steps:

1. Select a task/job to improve.
2. Get all the facts and information.
3. Create a process chart.
4. Question every aspect, list alternatives, and improve the necessary details.
5. Develop the preferred method.
6. Introduce it and evaluate the results.

What are the techniques and tools of work simplification?

Work simplification techniques include work distribution analysis, process analysis, motion analysis, work count, and layout studies. Work simplification tools include flow process charts, flow diagrams, and economy of movements.

What are some operational challenges for manufacturing engineers?

Some operational challenges include finding skilled labor with the right skills, increasing productivity, project management, improving internal production processes, inventory control, maintaining product and service quality, strengthening relationships, and responding to client demands.

What is ROI?

Return on investment (ROI) is a performance metric used to determine an investment's effectiveness. It aims to calculate the rate of return on a single investment explicitly relative to the investment cost. The profit (or return) of an investment is divided by the investment cost to determine the ROI.

For example, before deciding to purchase new manufacturing equipment, which can be a significant capital expenditure, we need to ensure that it is a profitable financial decision. It will have a high return on investment if it is a worthy purchase.

What are some different ways to increase ROI?

1. Increase sales in an appropriate manner. For instance, businesses may increase sales in specific regions during certain months instead of setting increased sales as a general goal.

Calculating the ROI involves figuring out the return on an investment over the cost of producing said investment.

2. Update marketing strategies by using digital marketing, i.e., posting advertisements through search engines, social media, etc. Manufacturers can reduce the costs associated with conventional marketing (including advertisements on TV, newspapers, etc.).
3. Reduce costs by modifying the design or packaging materials without losing the product's quality and negotiating with suppliers to get discounted prices.

What is the meaning of IIoT (Industrial Internet of Things)?

Industrial Internet of Things (IIoT) is a means of digital transformation in manufacturing. To gather crucial production data, IIoT utilizes a network of sensors and uses cloud computing to convert this data into useful information about operational activities' productivity. For example, IIoT manufacturing is used for monitoring operations remotely, predictive maintenance, smart asset management, etc.

What are the key advantages to implementing IIoT in a manufacturing environment?

- Helps to reduce the production lead time due to fast and more efficient production and supply chain operations
- Allows for mass customization by being a source of the real-time data needed for thoughtful forecasting, shop floor scheduling, and routing
- Facilitates monitoring of employee health conditions and hazardous operations that can lead to accidents

What is enterprise resource planning (ERP)?

Previously, businesses had separate software systems for every department, including accounting, finance, and HR, which worked independently without any interaction. With the introduction of ERP software, all these different processes came to the table to collaborate and create one fluid system. This system allows each department to see what the other is doing. As a result, accounting and HR can collaborate effectively for sales and customer relations.

What are the benefits of utilizing ERP in the manufacturing industry?

ERP for manufacturing is used to identify and organize the resource requirements of the entire organization.

- Allows access to the data needed under one system, which includes both operational and financial information
- Develops a reliable supply chain with increased visibility of stock as well as inventory and transportation management tools
- Increases customer satisfaction with one source of information about supply and demand
- Enables use of a multinational workforce and improves efficiency at multiple locations
- Generates a profit with detailed predictions based on real-time sales data plus new strategies to manage material requirements

What are the general responsibilities of a manufacturing engineer?

- Developing new products and techniques
- Procuring and installing equipment
- Repairing equipment
- Responding to machine and process breakdowns
- Analyzing industrial problems
- Making modifications to existing procedures in order to increase performance
- Managing technical and professional personnel
- Managing resources
- Managing statistical and financial reports
- Diagnosing faults
- Working closely with vendors, clients, and employees in research and development

What is the future scope of manufacturing engineering?

In the modern era, industrialization is gathering speed. Specialized people who can take over the manufacturing departments of industries are very much in demand. From an economic perspective, manufacturing engineering is a well-regarded job with a strong career outlook and decent pay.

The rapid advancement in manufacturing technology, particularly artificial intelligence, advanced robotics, and the Industrial Internet of Things, mostly referred to as Industry 4.0 technologies, are expected to transform the manufacturing ecosystem significantly.

Artificial intelligence, biomechatronics, autonomous cars, and nanotechnology are some of the fields where mechanical engineers are anticipated to be in high demand. There is a rising need for individuals with education in manufacturing engineering. By 2029, the U.S. is expected to add 30,000 more jobs in this industry.

CAREERS

What are some career pathways in manufacturing engineering?

People interested in manufacturing engineering careers will study materials that are related to the technologies necessary for the planning, managing, and processing of materials into intermediate or final products; designing and developing production planning and control; maintaining physical systems; and manufacturing process engineering. An individual can earn different certificates and degrees in order to be ready for a manufacturing engineering career. Potential careers in manufacturing engineering include quality technician, 3D CAD modeling, product design, and production planning and scheduling.

What certifications are available for manufacturing engineering?

- Professional Engineer (PE) certification and licensure
- Certified Manufacturing Technologist (CMfgT)
- Certified Manufacturing Engineer (CMfgE)
- Certified Engineering Manager (CEM)

Civil and Architectural Engineering

What are the biggest challenges that civil engineers and architects face?

Some of the challenges that civil engineers and architects face are as follows:

- Safety and well-being
- Compliance with different regulatory bodies
- Environmental issues such as natural disasters (hurricanes, tsunamis, and earthquakes), contamination and sediment runoff, extreme weather conditions, etc.
- Challenges related to contracts
- Monitoring the health of infrastructures
- Improving construction productivity
- Enhancing construction site safety
- Dealing with labor shortages
- Frustrations that construction companies are slow at adopting new technologies such as the use of drones, 3D printing, wearables, and autonomous vehicles

What is the difference between civil engineering and architecture?

Civil engineering focuses on a design's structural aspects, ensuring that a building can withstand both normal and extreme conditions, whereas architecture focuses on design aesthetics, look, and functionality. Also, architects are more engaged in the development stage, while civil engineers are in full control of all phases of construction.

CIVIL ENGINEERING

What is civil engineering?

Civil engineering is one of the oldest engineering disciplines. It is concerned with developing and designing the infrastructure for a civilization. Throughout time, civil engineering has formed a basis of knowledge from which engineers make decisions. This knowledge includes an understanding of the physical and natural sciences, mathematics, computational methods, economics, and project management. Civil engineers have several areas of specialization:

- Structural engineering
- Geotechnical engineering
- Water resources engineering
- Construction engineering
- Geotechnical engineering
- Transportation engineering

What is the role of geotechnical engineers?

Geotechnical engineering is a civil engineering specialty that entails researching and comprehending what lies under the ground's surface. Geotechnical engineers ensure that the foundations of constructed objects, such as highways, houses, runways, and dams, are stable. They emphasize how civil engineering systems, such as buildings and tunnels, communicate with the earth. Geotechnical engineering applications include military, mining, petroleum, coastal engineering, and offshore construction. Slopes, retaining walls, and tunnels are also designed and planned by geotechnical engineers.

Geotechnical engineers have the key job of ensuring that a construction site is geologically stable for the project that will be built there. This is why they need a solid knowledge of another science—geology—in order to perform their jobs.

What are some examples of civil engineering from early history?

Many examples of civil engineering can be found all around the world. Ancient civilization accomplished many engineering feats that are still difficult for us to understand and replicate. Some of the earliest civil engineering accomplishments occurred on the continent of Africa among ancient Egyptians and Iraqis between 2000 and 4000 B.C.E. Like civil engineering today, the structures that these ancient engineers built required knowledge of mathematics and many tools to accomplish, and it is important to note that the tools were much more simple than the ones we use today. Some examples are as follows:

- Egyptian Pyramids (c. 2700–2500 B.C.E.)
- Qanat water management system (c. 1000 B.C.E.)
- Parthenon, ancient Greece (447–438 B.C.E.)
- Great Wall of China (220 B.C.E.)
- Machu Picchu, Peru (c. 1450)

What do civil engineers do?

Modern civil engineers are involved in the entire life cycle of a project: project exploration, planning, construction, demolition, conception, design, op-

Civil engineers can be involved in all aspects of building projects, from exploration of sites to planning, construction, and demolition. They work on everything from bridges and tunnels to roads, buildings, airports, and water and sewage treatment plants.

timization, monitoring progress, and supervising and maintaining infrastructure projects. Their work is most noticeable on roads, buildings, airports, tunnels, dams, bridges, and systems for water supply and sewage treatment.

Other civil engineering tasks include analyzing plans, survey reports, and project designs; risk analysis; verifying that projects comply with various regulations; analyzing the adequacy and strength of building foundations; and cost estimation of building materials, equipment, labor, etc.

What skills do civil engineers need?

Civil engineers must have:

- An understanding and working knowledge of science
- Mathematical skills that include understanding algebra, calculus, trigonometry, and geometry
- An ability to resolve issues in design and development
- An ability to carry out technical and feasibility evaluations, particularly site investigations
- Experience designing comprehensive plans using a variety of computer tools
- Knowledge of project resource management, including budgeting
- Knowledge about how to oversee and evaluate the effects of soil tests in order to assess a foundation's suitability and strength

- Experimental design and analysis skills in order to analyze construction materials for use in specific applications such as concrete, wood, asphalt, or steel
- Time-management skills since large structural projects require efficient planning and scheduling if they are to be completed on schedule and on budget

How does a civil engineer model and prototype?

This work includes the use of computers and special software to create designs and models. Commonly used software includes computer-assisted design, or CAD, programs that help in the design process, store and retrieve information, and avoid design flaws. Furthermore, diazo copiers and special laser printers are used to produce blueprints, and chipboard and foam are used for making design models.

What are some famous civil engineering structures?

Civil engineers are fond of saying that it's architects who put designs on paper, but it's engineers who actually get things built. Some famous civil engineering structures include the world's tallest building, the towering Burj Khalifa in Dubai; the Chunnel, the 31-mile-long tunnel beneath the English Channel; the Golden Gate Bridge, which is a mile-long suspension bridge connecting the city of San Francisco to Marin County; Hoover Dam; and the Brooklyn Bridge, which is the first steel-wire suspension bridge in the world.

Who was the first civil engineer?

Civil engineering accomplishments occurred long before any person had the title civil engineer. John Smeaton (1724–1792) designed and constructed the Eddystone Lighthouse during the years 1756–1759 and later formed a group called the Smeatonian Society of Civil Engineers. The Eddystone tower is quite the feat. It was the third attempt by an engineer to build a lighthouse in that location. It was a known dangerous and very rocky location near Plymouth, England. Unlike earlier lighthouse designs built from wood, it was built entirely of stone. This was the first of its kind, and designers and builders were influenced by this design approach.

A physicist and mechanical engineer of note in Britain, John Smeaton is considered the "father of civil engineering."

Some aspects of his design approach that were novel (but made popular after his successful construction) were interlocking blocks of stone (the dovetailed joints in stone) and using quick-drying cement that could withstand the salty seawater (as opposed to wood, which would not be able to withstand water and storms). Forty-seven years later, the world's first engineering society, the Institution of Civil Engineers, was founded.

What is the goal of marine civil engineering?

Marine civil engineers primarily work to plan, calculate, design, and build structures in coastal zones, ports, and the Arctic. The major specializations of marine civil engineering include Arctic offshore and coastal technology, offshore wind technology, and coastal technology. These domains cover topics such as coastal and port engineering, Arctic technology, design of wind turbines, risers, and pipelines, etc.

What is the role of civil engineers in the Air Force?

Civil engineers maintain the buildings and facilities of the Air Force and make them battle ready. Also, they are in charge of developing and implementing solutions to complicated challenges in order to keep the infrastructures and utilities operational. Furthermore, these offer invaluable knowledge along with a wide spectrum of experience to thousands of structures throughout the world from drawing, surveying, and designing explosive ordnance disposal to disaster preparedness.

What is the role of civil engineers in the space program?

- The command centers for space programs are designed and built by civil engineers. This includes several departments, runways, warehouses, and hangers.
- They assist in the development of launch platforms that are used to launch spacecraft into and beyond Earth's atmosphere.
- They also collaborate on space station architecture with structural engineers, astrophysicists, and scientists.
- They can help build space elevators, which are long cables that are placed around the equator on Earth and extend out into space to transport payloads into low Earth orbit and back.

What is the role of civil engineers in stormwater management?

Civil engineers assist in stormwater management in multiple ways, including retention ponds, box culverts, flood control structure and design, and reservoir building.

How does a civil engineer design water slides for theme parks?

A civil engineer creates a pumping system that can provide exactly the right amount of water to the flume. The water slide would not function without this mechanism. A civil engineer also constructs the water slide to resist the flowing water, the weight of the people on it, and even the wind force blowing on it.

ARCHITECTS

What is the role of an architect?

Architects build plans for new buildings, renovations, and redevelopment programs. They design buildings that are operational, durable, sustainable, and aesthetically appealing using their expert architectural experience and high-level drawing skills. Thus, they are responsible for project planning and design as well as the aesthetic style of infrastructures and buildings.

What types of tools and equipment do architects use?

Design tools include tools for project designs that are created on paper. This requires the use of planes, triangles, protractors, architect's scales, templates to draw curves and circles of different sizes, special arm-and-track drafting machines, and a drafting desk.

Some basic tools include:

- Architect's scale ruler
- Desk lamp
- Drafting board
- T-square

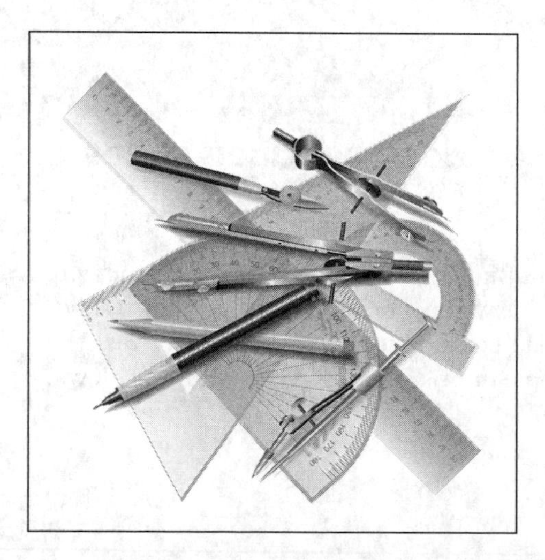

Despite the advent of computers that can aid in design, many architects still use tools like these to create drafts and blueprints on paper.

DID YOU KNOW?

What is an architectural drawing?

An architectural drawing is a technical representation of a structure (or construction project) that is utilized for a variety of reasons, including developing a design concept into a comprehensive proposal, communicating ideas and concepts, attracting customers to a design's merits, and aiding building contractors in construction. The drawing is based on detailed designs, using a set of rules that include specific views, sheet sizes, measurement units and scales, annotation, and cross-referencing. Thus, these are a collection of drawings, diagrams, and blueprints used to design, create, and record structures.

- Drafting chair
- Compass
- Lining pens
- Mechanical pencils
- Electric eraser
- Tracing paper
- Cutting mat
- Paper trimmer
- Precision-cutting knives
- Protective gloves
- Adjustable triangle
- Engineer's scale ruler
- Metal rulers
- Drafting brush
- Measuring tape
- French curve
- Laptop
- Architectural software

What are the responsibilities of a construction manager?

- Coordinate and direct construction workers, contractors, and subcontractors and maintain work quality

- Cost management
- Oversee both on-site and off-site projects to ensure that all building and safety regulations are followed
- Select machines, supplies, and machinery and keep track of inventory
- Execute according to the terms of the contract
- Constantly check on the status of the project

What are the skills required to become an architect?

- Strong mathematical and scientific skills: This includes geometry, algebra, engineering, programming, and physics, which are necessary for estimations, budgeting, etc.; for example, the task of converting scales from blueprints
- Design skills: This includes conceptualization, creative thinking, and industrial design along with innovation, drawing, and drafting; also includes designing concepts as well as design to delivery services
- Analytical and problem-solving skills
- Team-building skills
- Communication skills: This involves maintaining client relationships, collaborations, interpersonal skills, and coordination
- Experience with computers and software: This includes architectural rendering programs, AutoCAD and other CAD software programs, Model Maker, Revit, etc.

Of course, design skills are essential for being an architect, which also entails a strong knowledge of mathematics, physics, and engineering.

ROADS

What is the role of transportation engineers?

Transportation engineers design, construct, run, and maintain daily structures like streets and highways as well as larger projects like airports, ship terminals, mass transit systems, and harbors. It is the application of science and its concepts to the planning, functional design, service, and management of facilities for any mode of transportation to ensure the safe, reliable, rapid, secure, accessible, cost-effective, and environmentally friendly transportation/movement of goods and people.

What is the role of a road engineer?

Road engineers are responsible for the following duties:

- Overseeing and managing road construction, maintenance, and operations on the construction site
- Examining the soil for quality and strength at the selected construction site
- Inspecting everything, including the concrete strength quality
- Analyzing traffic influx, trends, and environmental dangers
- Creating roadways that are both safe and environmentally friendly
- Overseeing communication, approvals, quality checks, work permits, and other important issues related to the construction site and keeping those outside the company in the loop
- Examining carbon emissions and waste generated from road construction and determining how to properly dispose of them

What types of road construction are both durable and cost effective?

The types of road construction that are both durable and cost effective are as follows:

- Whitetopping roads
- Polymer fiber-reinforced concrete roads
- Bituminous roads

- Composite pavement roads
- Gravel roads

What are whitetopping roads? What are their advantages?

The process of overlaying an existing asphalt pavement with a layer of Portland cement concrete is known as whitetopping. The thickness of the concrete layer and its adherence to the asphalt base determine the kind of whitetopping. The main goal of an overlay is to either restore or enhance the load-carrying capacity of the existing pavement or both. Thus, it restores the rideability of existing pavements that have experienced rutting, deformation, texture loss, and deterioration.

The advantages of whitetopping roads are as follows:

- Strengthens the structural integrity of existing pavements
- Despite the initial cost being slightly more than that of bitumen roads, the life-cycle cost is significantly less than that of both asphalt and concrete roads
- Much quicker than creating concrete roads, with a turnaround period of only 14 days
- Increases light reflection and improves visibility and commuter safety at night, thus resulting in energy savings of up to 20–30 percent
- Reduces pavement deflection, resulting in up to 10–15 percent lower vehicle fuel consumption and, as a result, lower emissions
- Reduces the braking distance of vehicles, making them safer on both dry and wet surfaces

Fiber-reinforced concrete has many advantages over traditional concrete, including better strength, road safety, and drainage while reducing costs and installation time.

- Reduces the urban heat island effect by absorbing less heat, resulting in decreased energy usage for air conditioning in cities
- Provides excellent surface drainage
- 100 percent recyclable

What are polymer fiber-reinforced concrete roads? What are their advantages?

Polymer fiber-reinforced concrete roads are a composite material made up of asbestos, glass, plastic, carbon, or steel fibers mixed with cement paste, mortar, or concrete. When compared to nonreinforced concrete, fiber-reinforced concrete has a higher tensile strength. Polymeric fibers are currently typically adopted due to their property of no corrosion risk and their cost effectiveness. Polyester or polypropylene are the most common polymeric fibers used in these types of works. Common applications include highways, local streets, parking lots, sidewalks, etc.

The advantages of polymer fiber-reinforced concrete roads are as follows:

- Increases the concrete's durability and impact strength while reducing crack growth
- Employed in civil constructions to prevent corrosion to the greatest extent possible
- Utilized in bridges to prevent catastrophic collapses
- Improves fatigue strength and resistance against freezing and thawing
- Provides additional structural capacity in ultrathin whitetopping pavements

What are bituminous roads? What are their advantages?

Bituminous roads have a surface made of bituminous materials, which are also known as asphalt. It is a viscous, sticky, black liquid derived from natural reserves such as crude petroleum. Asphalt bitumen is a binding, organic substance that is utilized in road building because it is simple to make, reusable, nontoxic, and has a high binding ability. It is mostly used on low-traffic roads and as a sealing coat to renew an asphalt concrete pavement. In bituminous roads, aggregate is placed over a sprayed-on asphalt emulsion or cut-back asphalt cement where it is rolled into the asphalt, usually with a rubber-tired roller.

The advantages of bituminous roads are as follows:

- Provides a smooth driving experience since it has no joints
- Emits less noise than concrete pavements
- Retains its smoothness since the road wear and tear is minimal
- Degradation and failure is a very gradual process as compared to concrete roads
- Relatively lower cost and overall maintenance cost than concrete pavement
- Repairs are completed quickly so that the road may be reopened to traffic
- Unaffected by de-icing materials and resistant to high temperatures from melting

What are composite pavement roads? What are their advantages?

A Portland cement concrete sublayer is combined with an asphalt layer in composite pavements. They're more commonly utilized to repair existing roads than to build new ones. Asphalt overlays are sometimes used to repair damaged concrete to create a smooth wearing surface.

The advantages of composite pavement roads are as follows:

- Safe, smooth, quiet, robust, and durable
- Requires less maintenance than a conventional pavement system
- Long life
- Minimal fatigue cracking
- Lower life cycle costs

What are gravel roads? What are their advantages?

A gravel road is an unpaved road with a gravel surface that has been brought to the site from a quarry or stream bed. Stone, sand, and fines are the three types of aggregate that make up gravel. The optimal blend of these three groups varies considerably depending on the material's intended function. An excellent surface material for a gravel road, for example, would require more material to pass through a #200 sieve than a good base material.

The construction of gravel roads consists of a path of the road that would be excavated down many feet in the beginning, and French drains may or may not be installed depending on local circumstances. Large stones are then laid and compressed, followed by successive layers of smaller stones, until the road surface is made up entirely of small stones that are compacted into a firm, durable surface.

The advantages of gravel roads are as follows:

- Cheaper to design and build
- Easier to maintain since they require less equipment and may require less operator expertise
- Generate slower speeds than asphalt roads

BUILDINGS

What tools are needed for construction project management?

Necessary tools include a digital camera for documenting construction progress and design changes; management software and word-processing programs; and email and a cell phone for communication, negotiation, and management of vendor contracts and tracking of construction progress.

What is the foundation of a building? What are its functions?

The lowest portion of a building's construction is the foundation, which lies on the ground. Any foundation's purpose is to securely support and carry the building's combined dead, live, and wind loads to the ground without causing damage to any portion of the structure.

The types of foundations used in buildings are as follows:

1. Shallow Foundation:
    - Individual footing (pad foundation): This foundation is built for a single column, and the individual footings are square or rectangular. It is utilized when the structure's loads are supported by the columns.
    - Combined footing: This is constructed when two or more columns are close enough and their individual footings overlap each other.

- Strip foundation: These have a broader base than standard, load-bearing wall foundations. This broader base distributes the weight of the building structure across a larger area, improving stability.
- Raft or mat foundation: These foundations are laid throughout the structure to sustain the tremendous structural loads imposed by columns and walls.

2. Deep Foundation:

- Pile foundation: This is utilized to transmit enormous loads from a building to a hard, rock strata much below the ground level.
- Drilled shafts or caissons: These are high-capacity, cast-in-situ foundations that are similar to pile foundations. They resist structural stresses using shaft resistance, toe resistance, or a combination of the two.

When is pile foundation typically used in construction?

When soil conditions are insufficient to handle a project's design loads, pile foundations are frequently used. When dealing with compressible or weak topsoil layers, soil erosion, a soil-bearing capacity of less than 24 kilonewtons per cubic meter (24kn/m^3) or horizontal stresses we utilize a pile foundation.

What is the purpose of using columns in a building, and how much steel do columns require?

Columns are the vertical components of a structure that sustain a structural load that is distributed throughout the structure via beams. After the load

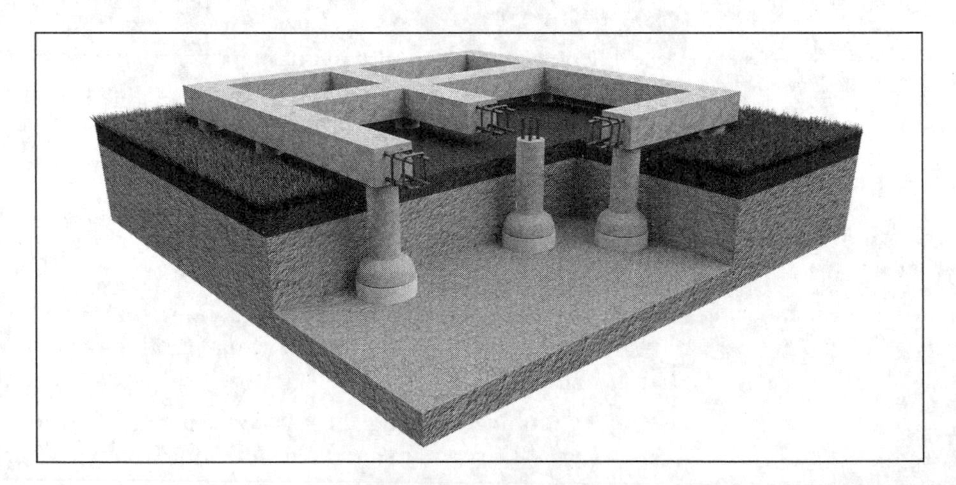

Figure 12: A drilled belled foundation (top) vs. a screw-piles foundation (bottom) are two approaches to creating deep foundations that will support a large building such as a skyscraper.

is transferred from the column to the footing, the weight is transferred from the footing to the land. The maximum longitudinal steel is 6 percent without lapping and 4 percent with lapping, according to clause 26.5.3.1 of IS 456-2000, whereas the lowest longitudinal steel is 0.8 percent.

What kinds of metals/alloys are used in construction?

Titanium, stainless steel, aluminum, copper/copper tubing, lead, and iron are the most common alloys used in construction. Each alloy has a specific application and provides the strength and toughness needed in construction. For example: Titanium is a solid, lightweight metal with excellent corrosion resistance. It is mostly used in air conditioning and heating systems as well as plumbing and roofing. Furthermore, copper is a malleable and ductile metal with a high thermal and electrical conductivity. It is used in electrical wiring, heating, rainwater systems, etc.

What is a TMT bar in steel?

TMT is the abbreviation for thermo-mechanically treated. TMT bars have a hard exterior and a softer core and are produced from carbon steel. The bars are extremely strong and resistant to corrosion. Other characteristics include:

- Flexibility
- Weldability
- Fire resistance
- Sturdy build
- Corrosion resistance
- High dimension tolerance

Does water damage concrete?

Water is one of the main ingredients used in the concrete mixing and curing process. In the mixing process, water is used with cement and sand. In the curing process, the newly poured concrete undergoes "moist curing," which means that it is frequently hosed down with water. This allows for the moisture in the concrete to cure more slowly, which is the most preferred method. This will not harm or destroy plain concrete; however, water can harm reinforced concrete (RC). This is due to the possibility of water seeping into the concrete and corroding the steel reinforcement within.

What is curing?

Curing is the process of controlling the evaporation process (moisture loss) from concrete until it meets the desired design properties.

When does concrete achieve full strength?

Although concrete hardens soon after it is poured, what the full strength of concrete is depends on its intended purpose. The construction industry commonly recognizes that a period of 28 days is necessary to wait before concrete achieves its full strength. For typical weather conditions, according to code of practice IS 456-2000, concrete should not be cured for less than 7 days. When concrete is cured for 3–7 days, it achieves 50 percent of its intended strength. After 14 days, it achieves 75 percent of its compressive strength, and after 28 days, it achieves 90 percent of its compressive strength. Strength progressively grows over time. These curing periods are accepted standards, but the actual curing time varies based on mix designs.

What are different methods of curing?

Curing is done in a variety of ways, the most frequently used of which are:

- Shading concrete work
- Using gunny bags to cover the concrete surface
- Water spraying or sprinkling
- Ponding
- Using curing membranes
- Using steam

What are some basic loads on structure?

- Structure self-weight, i.e., slabs, beams, columns, walls, etc.
- Furniture, equipment, machinery, and other live loads

DID YOU KNOW?

What are the various tests used to examine construction materials?

Soil tests include tests for core cutting, soil compaction, sand replacement, and consolidation. Concrete tests include tests for slump, compression, split tensile strength, and soundness. Bitumen tests include tests for ductility, softening point, and gravity penetration.

- Wind loads
- Earthquake/seismic loads
- Snow loads
- Hurricane loads

What is the role of construction engineers?

Construction engineers oversee the planning and execution of construction projects, ensuring that they are completed on time and according to requirements. Usually, these engineers are in charge of the design and safety of temporary structures that are used during construction. They may also be in charge of a project's budgeting, time management, and communications. Infrastructures such as highways, tunnels, bridges, airports, railroads, buildings, dams, utilities, and other projects are designed, planned, built, and managed by construction engineers.

What is the role of structural engineers?

Structural engineering deals with creating well-designed and efficient structures, which include buildings, bridges, tunnels, machines, and vehicles. Consider regions that are more prone to earthquakes. Structural engineers analyze and design buildings in these regions that can withstand earthquakes. They consider the structural integrity by analyzing the constraints. Common constraints that structural engineers are concerned with include stiffness, strength, and tension.

What are different types of construction?

Over time, engineers have been able to develop more diverse types of structures because of advancements made in computational methods. For example, in order to determine how a structure would behave under a specific stress in the past, civil engineers would use free-body diagrams and mathematical equations on statics and dynamics from mechanical engineering. Unfortunately, because advanced computing and programming were not available before the 1950s, civil engineers were limited in what they could safely build with the same levels of confidence that can now be had with advanced software.

The construction industry typically groups constructions and buildings into one of five specific types:

Type 1. Fire-resistive construction

Type 2. Noncombustible construction

Type 3. Ordinary construction

Type 4. Heavy-timber construction

Type 5. Wood-frame construction

What is fire-resistive construction?

This is a construction in which the exterior (walls, floors, and roof) is made of noncombustible materials such as steel or concrete. These materials

It's not only the main materials that can protect a building by fire; this worker is applying a special kind of fire-resistant paint at a construction site.

have a fire resistance rating of at least two hours. They can resist fire for extended periods of time or contain the floor or space of origin and allow for evacuation. High-rise and mid-rise buildings are commonly constructed with fire-resistive, noncombustible materials.

What is noncombustible construction?

This structure type is commonly found in newer school buildings and commercial buildings. The exterior walls, structural frame, and flooring/roofing have a fire-resistance rating that is less than the fire-resistive construction rating. Protected noncombustible structures can resist and contain fires for one hour. Unprotected noncombustible constructions, which are used in commercial buildings, are built with noncombustible materials that have no fire resistance.

What is ordinary construction?

Ordinary (Type III) construction can be divided into two types. These structures are also called brick-and-joist designs. All or some of the structural elements of the structures' interiors may have combustible properties. A structure would usually be two or three floors tall, with a maximum height of six stories.

Protected combustible structures have brick or block walls and wooded flooring/roofing. The walls have a two-hour fire resistance rating, and the flooring and ceiling have one hour.

Structures in warehouse districts are typically unprotected combustible structures. They also have brick or block exterior walls; however, the wooden roofing, ceiling, and flooring are not protected against fire.

What is wood-frame construction?

Wood-frame structures are the most combustible of all the construction forms. These permit the use of wood for combustible exterior walls and combustible interiors (structural supports, walls, floors, and roofs), and they include two subgroups: protected wood-frame construction and unprotected wood-frame construction. Protected wood-frame structures are commonly used in the construction of newer apartment buildings. In these structures, no wood is exposed, and they have a fire rating of one hour. However, unprotected wood frames, like those used in garages, frequently have exposed wood, which means that they are not fire resistant.

DID YOU KNOW?

What is heavy-timber construction?

The National Fire Protection Association (NFPA) defines heavy-timber construction as a system having main framing members measuring no less than 8 inches by 8 inches with exterior walls that are made of a noncombustible material. This was previously known as a mill construction because textile factories and paper mills were the first to use this type of construction. The flooring must be a minimum of 3 inches thick. Due to the thickness and type of wood used to make these structures, it takes a longer amount of time for the building to catch fire, and they do not collapse easily due to the structural mass. However, if one of these structures catches on fire, it would take a large amount of water to extinguish it.

What are the different types of cement?

Cement is mainly classified into two categories that are determined based on how the cement hardens. Hydraulic cement includes those that harden with the presence of water and forms a water-resistant product. Nonhydraulic cement uses carbon dioxide from the air to harden, so it requires dry environmental conditions. Different types of cement are used for various construction needs. Each type has a composition that meets the specific need of the project.

- Ordinary Portland cement (OPC): The most commonly used cement form and suitable for all types of general concrete construction
- Portland pozzolana cement (PPC): Has a high resistance to multiple chemical attacks on concrete and is used in marine infrastructure, sewage works, and underwater concrete laying such as bridges, piers, dams, mass concrete works, etc.
- Rapid-hardening cement: Has a higher lime content, higher C3S content, and finer grinding, resulting in a stronger early strength growth than OPC; formwork can be removed sooner, increasing construction speed while lowering construction costs by reducing formwork costs; used in prefabricated concrete buildings and road construction
- Quick-setting cement: Sets earlier than rapid-hardening cement, with the rate of gain of strength similar to OPC; used for concreting in static or flowing water and where work must be done in a short time period
- Low-heat cement: Ideal for massive concrete construction such as gravity dams, as the low heat of hydration prevents the concrete from crack-

ing due to heat; better resistance to sulfates, less reactive, and longer initial setting time than OPC

- Sulfate-resisting cement: Used in the building of foundations where the soil has a high sulfate content in order to minimize the chance of sulfate attack on concrete; commonly used in areas like canal linings, culverts, retaining walls, and siphons where a lot of sulfate action occurs due to the water and soil
- Blast furnace slag cement: Used for works where cost is a major concern
- High-alumina cement: Has a higher compressive strength and is more workable than ordinary Portland cement, making it ideal for projects involving high temperatures, frost, or acidic action
- White cement: A white variety of ordinary Portland cement; more expensive and used for construction uses such as precast curtain wall and facing panels and terrazzo surfaces; also used in internal and exterior decorative work such as building facing slabs, floorings, ornamental concrete items, garden paths, etc.
- Colored cement: Made by combining 5–10 percent mineral pigments with regular cement; commonly used for decorative floor work
- Air-entraining cement: Made by grinding clinker with additives like resins, glues, sodium sulfates, etc.; particularly well suited for enhancing workability with a lower water–cement ratio and improving concrete's frost resistance
- Expansive cement: Stretches slightly with time but does not shrink before or after the hardening process; commonly used for anchor bolts and prestressed concrete ducts
- Hydrographic cement: Made by combining chemicals that repel water and has a high workability and strength; can repel water and is unaffected by rain or monsoons; commonly used for constructing dams, water pipes, spillways, water retention systems, and other water structures

How is concrete produced?

A chemically inert mineral aggregate (usually sand, gravel, or crushed stone), a binder (natural or synthetic cement), chemical additives, and water are combined to make concrete, a hardened building substance. Concrete dries to a stonelike quality, making it suitable for building sidewalks, bridges, and other structures. Thus, it is a mixture of fine and coarse aggregates that are bound together with a cement paste that hardens over time.

DID YOU KNOW?

What are the optimal initial and final setting times of a cement mix?

For almost all types of cement, the initial setting time for an adequate cement mix is approximately 30 minutes. It might take up to 90 minutes for masonry cement. The ideal cement mix should be set in no more than 10 hours. Furthermore, for masonry cement, it should not take more than 24 hours.

What is cement segregation?

The separation of cement and sand from aggregate is known as segregation. This happens when the water–cement ratio is improper and the concrete is poured from a height of more than 1.5 meters.

At what temperature do the ingredients of cement burn?

Cement ingredients burn at 1,400 degrees Celsius.

What is honeycombing?

The term "honeycomb" refers to voids in concrete created by mortar failing to fill the gaps between coarse aggregate particles. Honeycombing may occur when unexpected vibrations occur during the concrete pour, when the coverage of reinforcing bars is limited, when a stiff mix concrete is used, or when a higher percentage of larger-sized aggregate is used in a concrete mixture.

What is guniting?

This is a process that was developed in the United States that is used in construction for the purposes of slop stabilization, rehabilitation of retaining walls, and in concrete repair. It is a procedure in which a 1:3 mixture of cement

A worker sprays gunite over rebar, a method that works very well to produce curved and complex surfaces such as swimming pools or zoo exhibits.

and sand is sprayed on a concrete surface with a cement gun at a pressure of 2–3 kg/cm2. The mix used in the guniting is a cement–mortar mix. Dry cement and sand are mixed in a bin and stored in a cement gun. At the time of application, the dry mix is injected with high pressure and high velocity and the water is added at the nozzle just before the mixture blasts out of the gun. The benefits of guniting are that this method can be used on structures that are spherically shaped or curved, the mix can be prepared quickly, and it is the best choice when a possibility exists that the work could be stopped or interrupted (less material loss because of the dry-mix approach).

What are the various tests for checking brick quality?

To determine if a brick is suitable for building work, it is subjected to the following tests:

- Crushing strength test: Bricks should have a minimum crushing strength of 105 kg/cm^2. Crushing measuring machines may be used to determine the crushing strength of bricks. Crushing force is applied to the brick during the measuring operation after it is attached to the crushing testing unit.

- Water absorption test: When a brick is dry, it is weighed. Afterward, it is submerged in water for 24 hours. Then, it is weighed again, and the difference in weight shows how much water the brick has consumed. In any case, it should not be more than 20 percent of the dry brick weight (particularly for first class).
- Presence of soluble salts (efflorescence test): The occurrence of soluble salts in brick is determined by immersing it in water for 24 hours. After that, it is removed and allowed to dry. The efflorescence is said to be mild when a white deposit covers about 10 percent of the surface and moderate when the white deposit covers about 50 percent of the surface. If white deposits cover more than half of the soil, the efflorescence is serious, and it should be eliminated.
- Hardness test: A scratch is attempted on the brick surface with the application of a fingernail in this test. The brick is considered sufficiently strong if no mark is left on the surface.
- Shape and size test: A brick should be of a regular size, with sharp edges and corners, and a true rectangular shape. In this analysis, 20 bricks of regular size (190 mm × 90 mm × 90 mm) are chosen at random and stacked lengthwise, widthwise, and heightwise. The measurements should be below the following permissible ranges for good-quality bricks:

Length: 3680–3920 mm

Width: 1740–1860 mm

Height: 1740–1860 mm

- Soundness test: Two bricks are taken and stacked on top of each other in this test. The bricks should not crack and should produce a clear, ringing sound.

You might take bricks for granted when you see them on a wall or as pavement, but they actually go through rigorous testing before they are used in construction.

- Structure test: The foundation of a brick is studied after it is broken. It should be homogeneous, compact, and free of flaws like holes and lumps.
- Color test: A good-quality brick should be a rich red, cherry, or copper color that is consistent throughout.
- Impact test: A brick is dropped from a height of 1 meter in this experiment. A high-quality brick can never crack. If the brick breaks, that means it has a poor impact value and should be discarded.

Earthquake Resistance

What measures are taken during the construction of buildings to prepare for earthquakes?

- Having a dedicated earthquake-protected area on the construction site
- Implementing and practicing earthquake-smart construction methods early in the construction process
- Practicing the technique "Drop, Cover, and Hold On" at least twice a year and at least once per new construction site
- Conducting earthquake drills to ensure that construction crew members are familiar with what to do in the event of an earthquake
- Watching for fires, which may be started by broken gas pipes, including at building sites, as well as by faulty electrical components or lines
- Keeping an earthquake emergency supply kit

What are some design elements that help structures withstand earthquakes?

- Building a flexible foundation: This entails building a structure on top of concrete, rubber, and lead, flexible pads. The isolators vibrate as the base shakes during an earthquake, but the foundation itself stays stable. This essentially absorbs seismic waves and prevents them from passing into a structure.
- Damping forces as a form of countermeasure: Shock absorbers help buildings slow down by reducing the magnitude of shockwaves. Vibrational-control devices and pendulum dampers are two methods of damping.

- Vibration shielding for structures: Buildings can deflect and reroute the energy from earthquakes using this process. This design involves the construction of a cloak of 100 concentric plastic and concrete rings and burying it at least 3 feet under the building's base. When seismic waves penetrate the rings, they are pushed to pass to the outer rings, where they can travel more easily. As a result, they are effectively channeled away from the structure, protecting the building from earthquakes.
- Strengthening the structure of the building: Buildings must redistribute the forces that pass across them during a seismic incident in order to avoid collapsing. Reinforcing a building requires the use of shear walls, cross braces, diaphragms, and moment-resisting frames.

What are deformation and deflection?

Deformation is a permanent change in the shape of a material caused by the application of force, heat, or other factors. When the force is withdrawn, it does not return to its former form. Deflection, on the other hand, is only temporary, and the deflected material will return to its original shape once the force that caused it is withdrawn.

Why do engineers use reinforcement in concrete?

The most common form of concrete used in building is reinforced concrete, which is a mix of concrete and reinforcement. Deformed steel bars are embedded into recently manufactured concrete at the point of casting to provide reinforcement. The aim of reinforcement is to give concrete more support where it's required. Steel contributes much of the tensile strength where concrete is under tension such as in beams and slabs. Also, it provides additional stiffness to beams.

What causes a building to collapse?

Buildings can collapse for many reasons. A typical explanation for a building collapse is a structural failure due to:

- Errors in structural design and drawings ranging from incomplete details (i.e., missing dimensions) to not following design codes
- Weakness of foundation

- Weakness of building materials
- Mistakes in mixing the materials (i.e., concrete)
- Heavier than expected load
- Weakness of the building

BRIDGES AND DAMS

How are dams built?

A dam is a structure built to serve as a barrier across a body of water such as a river. It can act as a plug to prevent water from flowing across it, or it may simply restrict the flow of water without completely blocking it. Both humans and animals build dams. Beavers are most recognized for their dam building. They use sticks and mud to restrict the flow of water and create wetlands to make homes for their families. This also influences the environment around them and attracts many different types of wildlife. Humans build dams for many purposes. One common purpose is to hold back water and create a reservoir to generate energy. Other purposes include irrigation and creating a drinking water supply. When civil engineers design dams, they must take into account the force of the water building behind the dam and ensure that they select the proper materials to maintain it under the stress.

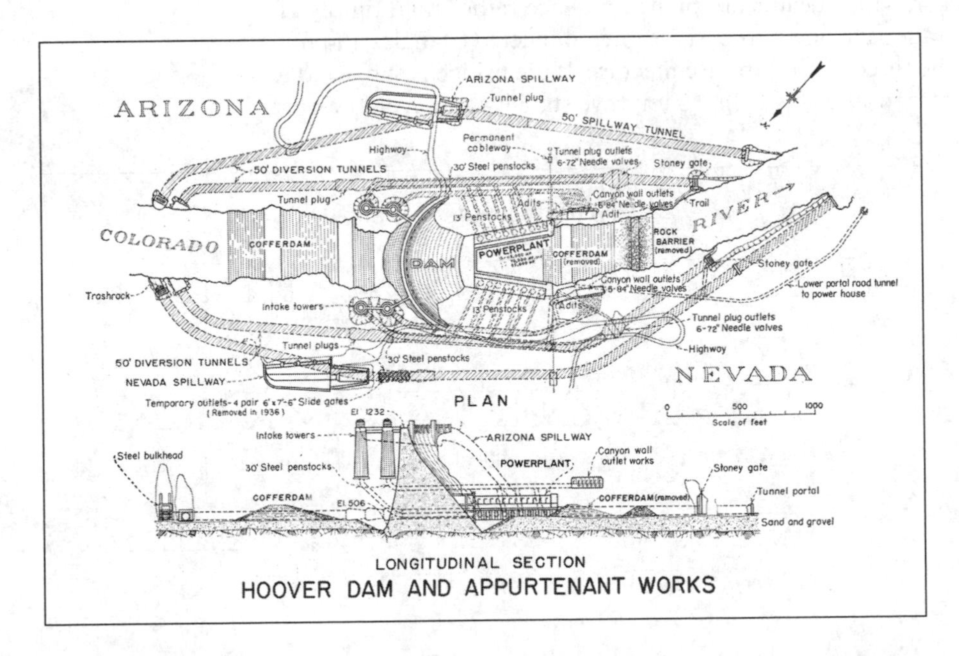

Figure 13: A modern dam is more than just a wall holding back a body of water. This overview of Hoover Dam shows the complex systems put in place, including diversion tunnels, intake towers, cofferdams, the powerplant, and the dam itself.

What measures are taken to make freeway bridges safe?

Transportation is growing increasingly important as more countries are developing and growing. Roads, bridges, and tunnels must be very safe and reliable. Here are some strategies that civil engineers use to ensure the safety of all drivers and pedestrians:

1. Safety edge: A 30-degree incline on the edge of the pavement prevents drivers from unknowingly drifting off the road
2. Corridor access management: Making decisions about how and when vehicles should enter and exit the roads; this occurs at the start of the design, and engineers may choose to restrict the number of access points to a location so as to reduce the number of opportunities drivers have to get into accidents with other vehicles
3. Enhanced delineator and friction for horizontal curve: Increasing friction at curves so that nighttime drivers are aware that the stretch of road is no longer straight

What is the strongest type of bridge?

The strongest kind of bridge is a truss bridge. It provides a network of support structures in which beams are pinned in place rather than rigidly affixed, allowing vibrations to flow through the interconnected triangles and uniformly distribute the force around the architecture. If the bridge is subjected to strong winds or heavy traffic, this helps to improve stability and avoid flexing.

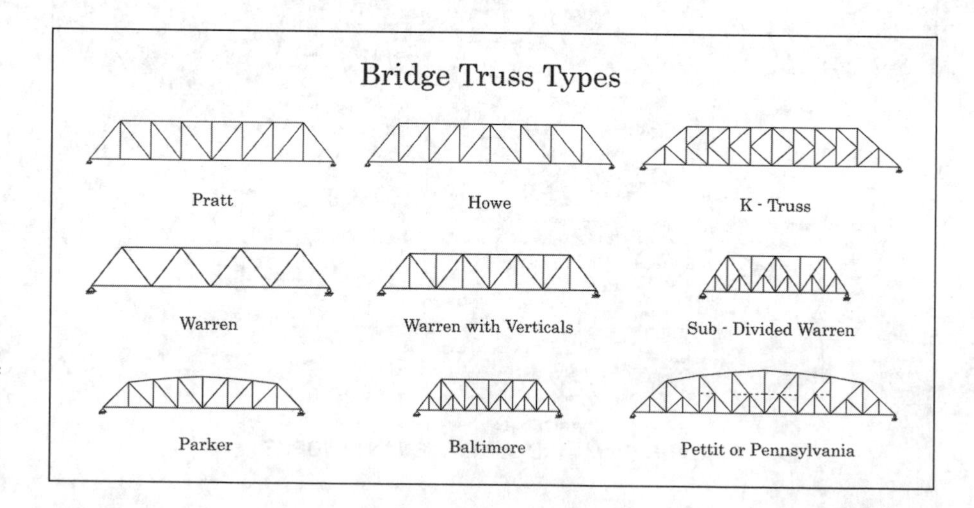

Figure 14: The strongest type of bridge is the truss bridge of which there are several varieties.

COMPUTERS AND SOFTWARE

What are some of the software programs that are useful for civil engineers and architects?

Some of the software programs that are useful for civil engineers and architects are as follows:

- AutoCAD Civil 3D
- HEC-HMS
- HEC-RAS
- Microsoft Project
- StormCAD
- SSA
- WaterCAD
- EPANET
- ArcGIS
- Bluebeam
- STAAD.Pro
- Primavera
- Revit Structure
- ETABS
- SAP 2000
- Microsoft Excel
- 3DS Max
- MX Road

SAFETY AND ENVIRONMENTAL CONCERNS

What is fatigue in construction, and how is it avoided?

Causes of fatigue in construction include extended hours for employees, physically and mentally demanding work, working in extreme temperatures, noise on the job site, inhaling fumes from hazardous materials and chemicals, professional conflicts with a supervisor, dealing with personal issues, etc.

Methods to avoid fatigue in construction include managing one's workload; using biometric sensors, such as the CAT smartband, to identify fatigue levels; developing corporate wellness programs; and keeping a positive culture with a supportive environment and commitment to safety.

What are the effects of weather on construction?

Certain extreme weather conditions are unsafe for construction and can cause bodily harm. Some concrete and masonry require specific temperatures and moisture levels for best performance. Exterior projects, such as roofing, can be hampered by heavy rain. Additionally, heavy rain or snow can flood the job site, making it difficult for construction. Excavation requires a thawed ground in warmer temperatures. Projects that include painting, waterproofing, and sealing must be done at certain temperatures. Due to seasonal weather restrictions, landscaping work is halted from late fall to early spring. Extreme weather, high altitudes, and wind can ruin some stored materials at construction sites. Mold can be caused by moisture, heat, and a lack of direct sunlight. These are some of the environmental characteristics that an engineer should consider when selecting materials and scheduling projects.

How does construction cause soil erosion?

Building and road construction churns up the earth and exposes soil to erosion. Native ecosystems, such as forests and grasslands, are cleared in some places, opening the soil to erosion. Root systems provide vegetation with the ability to keep soil in place in order to avoid erosion. Dirt is often in the air above construction sites, swept up by the wind and water erosion. Thus, when building machinery kills trees, soil erosion occurs.

What kind of inspections are done during construction?

Inspection of the building process is done in order to ensure that all materials and processes are in accordance with the design and requirements. On-site inspections are reported on a regular basis, including inspections for contractor operations. Engineers perform checks to see if the site meets the necessary requirements and specifications for quality. They also examine and discuss drawings and requirements for discrepancies in the project. Throughout this process, the engineer maintains records and inspection checklists. They also keep track of the project schedule and monitor construction material delivery and usage. Often, paint is applied during the construction process, so the engineer needs to inspect the paint and surface coatings to ensure that the products meet standards. A civil engineer can also perform material sampling and field testing on soil, concrete, asphalt, and other materials.

What are the different types of inspectors?

- Electrical inspectors examine electrical systems to ensure that they are in good working order and safe to use
- Mechanical inspectors inspect heating, ventilation, and cooling systems
- Plumbing inspectors test pipes, fixtures, water supplies, etc.
- Public works inspectors inspect sewers, dams, bridges, and roads

What types of risks are present on construction sites for workers?

Many types of risks are associated with working on a construction site. The most common cause of severe workplace injuries is falling from heights.

Inspections to ensure that construction sites meet codes to assure safety and quality are essential for any project.

DID YOU KNOW?

How does construction affect the environment?

The construction industry is responsible for 25–40 percent of global carbon emissions. This includes air and water pollution, waste that ends up in landfills, and fossil fuel combustion, which emits greenhouse gases and affects the atmosphere. Other factors include effects from high-energy usage in newly constructed buildings and negative effects on wildlife due to the clearing of vegetation and excavating.

Injuries are caused by ladders that are not properly secured, the lack of safety nets or guardrails, edges that aren't insulated, loose machinery on roofs or elevated walkways, slippery or wet surfaces, unchecked materials or facilities, uneven terrain or surfaces, loose cables, etc. Risks are also associated with constant movement of vehicles, machinery, and equipment. Noise levels on a construction site can be extremely dangerous. This can include high-powered tools, groundwork tools, heavy-duty vehicles, etc. These tools can also result in repetitive vibration. Furthermore, lifting, transporting, and treating material loads and equipment improperly on a regular basis will result in serious injuries.

Other risks include potential collapses due to working in an unstable area or weakened structures due to excavation. Lung cancer, asbestosis, and mesothelioma are all life-threatening respiratory diseases caused by asbestos, which can be found on construction sites. A major source of risk includes certain insulation types, old switchgears, ceiling tiles, wall plaster, etc. Direct contact with a power source or indirect contact with live equipment or wiring may result in injuries. This is mainly caused by damaged tools or equipment, inadequate wiring, or overloaded power boards. Finally, some chemicals used during construction can be harmful to both workers and the environment.

What are the steps taken by construction sites for waste management?

1. Using recyclable materials and products to minimize waste
2. Segregating different types of waste by using color-coded labels, bulk bags, etc.
3. Using incentives and rewards for employees to follow the steps of waste management

4. Performing deconstruction rather than demolition to reduce heavy pollution and waste generation, helping to preserve resources, minimize landfill waste, and create new employment opportunities

How are landfills constructed, and how do they affect the environment?

Modern landfills are well-engineered and maintained facilities for solid-waste disposal. Some of the negative effects caused by landfills are as follows:

- Land use: Large amounts of land are occupied by landfills. The amount of waste generated by modern societies is such that the amount of land required for landfills is a problem. This is a particularly pressing issue in densely populated areas where consumption rates are high. Land is being used for landfills at such a high rate that it is becoming an inconvenience to the locals. Recycling, reducing waste, and lowering usage rates are all part of the solution to this issue.
- Groundwater pollution (toxins): Many products dumped into landfills contain poisonous compounds that can damage soil and water, putting all types of life at risk. It's impossible to solve this problem because hazardous chemicals are often mixed in with millions of tons of less toxic trash.
- Air pollution (methane): When trash and waste are piled up, they begin to rot. When waste in landfills decomposes, methane is generated, which is much more potent than carbon dioxide. Methane is released into the atmosphere from the landfill, resulting in global warming.
- Smell: Garbage has an unpleasant odor. This is particularly a major issue for residents living near landfills.

Some landfills are being designed to be more ecologically sound. The sign at this one explains how methane gas released from garbage is being recycled for use in power generation. That's good engineering!

- Oceans: The "garbage patch" found in the Pacific Ocean is an example of a landfill's effect on oceans.
- Biodiversity effects: The creation of a landfill site results in the extinction of 30–300 species per hectare. Local species are also changing, with rats and crows replacing certain mammals and birds that eat garbage. Regardless of the length of the landfill site, vegetation modifications arise as certain plant species are displaced by others.
- Soil fertility effects: The combination of toxic substances and decaying organic material in the soil around a landfill site may have an effect on the soil quality of the surrounding areas. Local vegetation can cease to develop and be permanently altered, compounding the effects on biodiversity.
- Visual and health impacts: Disease becomes a problem as a result of the rise of the number of vermin around landfills; other negative health consequences such as birth defects, tumors, and respiratory diseases have also been attributed to landfill exposure.

Industrial Engineering

What is industrial engineering?

Industrial engineering is a field that focuses on products, systems, and people. Industrial engineers can design specific products to meet needs. They focus on quality, materials, and resources. They can design for people in cases where they observe how a product or process best meets the needs of individuals or how an individual might use a product with limited stress on their body. They might also redesign systems that people already use—for example, queues at the grocery store, the TSA line in the airport, or a theme park ride line—to ensure that people, animals, and products move through in an efficient manner.

Industrial engineering involves much more than what one might think given its name. Industrial engineering received its name during World War I because of the application of its methods to industrial workplaces such as plants. Early industrial engineers did time studies to examine how long it took to do specific activities. The engineers primarily focused on how to work more efficiently by optimizing complex processes and systems.

In more modern times, industrial engineering extends to so many different areas. Some areas of specialization associated with industrial engineering are manufacturing engineering, systems engineering, operations research, product and process engineering, supply chain management, engineering management, human factors engineering, logistic engineering, and many others.

DID YOU KNOW?

How many cars were built during this time?

Around 30 million cars were produced in the early days of automotive manufacturing. By 1930, one in every five people living in the United States had a vehicle.

Why was industrial engineering created?

Companies and government agencies in the United States wanted to be more competitive with other nations. The manufacturing industry creates many products that are used by people in the United States and abroad. Manufacturing could be very expensive and slow if the process of finding the materials, getting them to the facility, creating the products, and getting them to the customer is not refined. Many people from diverse backgrounds developed processes and solutions to help the United States become more competitive and more productive. These corporations built large facilities that built vehicles that would be driven all over the world. Engineers were needed to manage all the people, facilities (buildings), materials, machines, and manufacturing processes efficiently.

What role did cars (automotive vehicles) play in the beginnings of industrial engineering?

Around the beginning of the twentieth century, many small shops built cars. As more roads were paved, more individuals and families wanted to travel farther. Homes were more spread out, and people no longer lived in primarily urban centers. This increased the demand for cars, and large companies began to form as small shops were merged together.

Which automotive companies emerged at the inception of industrial engineering?

The large automotive corporations that were birthed from the small shops included Chrysler, Dodge, Ford, GM, Hudson, Packard, and Studebaker.

Why was industrial engineering important in the early twentieth century?

Because the demand was so high in the early twentieth century for products like cars and other manufactured products, industrial engineering was important to ensure that processes were always performed the same way and at the most efficient and productive pace.

PIONEERS OF INDUSTRIAL ENGINEERING

Who was Frederick Winslow Taylor?

In 1911, Frederick Winslow Taylor (1856–1915) published his book *Principles of Scientific Management*, which made popular the idea and term called Taylorism. Taylor was a pioneer who invested much of his time understanding and improving worker productivity. He carefully observed workers as they performed their duties and timed each step of their tasks. He documented the tasks and the details of the worker motion required to complete the tasks. Taylor was

American mechanical engineer Frederick Winslow Taylor pioneered methods for improving efficiency in industry.

an innovator/engineer who was well known in mechanical engineering circles in the late 1800s. He was known for finding the "one best way" to cut metal. He learned to vary metal-cutting components like the speed of cutting tools, how cutting tools were fed, and the shape of cutting tools in over 30,000 experiments. His success in optimizing this metal-cutting process led him to think big. The metal-cutting machine was only one aspect of a much larger manufacturing process. What if the whole manufacturing process could be optimized? This led him to develop and preach the gospel of efficiency, or Taylorism.

What are critiques of Taylorism?

Although his work was a great contribution to the field, critics often argued that scientific management, or Taylorism, reduced human workers to machine cogs. Workers were constantly under the watchful and critical eye of the timekeepers, and the well-being of the worker could easily be forgotten in favor of a highly optimized process.

What are the four principles of scientific management?

Taylor's four principles are as follows:

1. Replace working by "rule of thumb," or simple habits and common sense, with using the scientific method to study work and determine the most efficient way to perform specific tasks.
2. Rather than simply assigning workers to just any job, match workers to their jobs based on capability and motivation, then train them to work at maximum efficiency.
3. Monitor worker performance and provide instructions and supervision to ensure that they're using the most efficient ways of working.
4. Allocate the work between managers and workers so that the managers spend their time planning and training, allowing the workers to perform their tasks efficiently.

Frank and Lillian Gilbreth

Who were Frank and Lillian Gilbreth?

Frank Bunker Gilbreth Sr. (1868–1924) and Lillian Moller Gilbreth (1878–1972) were a married couple who had 12 children and are known as one of the

Husband-and-wife team Frank and Lillian Gilbreth made contributions in understanding motion and human factors in improving industrial efficiency.

greatest couples and contributors in the field of engineering and science. They developed motion studies that they used to identify the one best way to complete tasks and then replicated it to reduce the task to the smallest number of steps needed to complete it. They believed that this process could be applied to any job. They not only thought about how to make tasks more efficient, but they also extended their belief to how organizations and people are managed.

Where did the Gilbreths start their work?

The Gilbreths first applied their motion studies to the bricklaying industry and proposed improvements to the process of bricklaying, which addressed how often the bricklayer moved and turned, where the bricks and mortar were placed, and the introduction of a new scaffold.

What was Frank and Lillian Gilbreth's management theory?

The Gilbreths' organizational management theory included three main objectives.

1. Reduce the number of motions in a task.
2. Focus on the incremental study of motions and time.
3. Increase efficiency in order to increase profit and worker satisfaction.

DID YOU KNOW?

Did the Gilbreths really study at home?

Yes. After Frank Gilbreth died in 1924, Lillian Gilbreth continued to extend their motion studies to tasks at home and made improvements to the way homes were designed to help people with disabilities.

What is a time study?

One of the best ways to evaluate human behavior and machine interaction is to complete a time study. A time–motion study requires an industrial engineer to examine repetitive tasks by breaking the task into (1) small, simple steps, (2) observing each sequence of steps with careful attention to wasteful motions, and (3) measuring the amount of time it took for the worker to complete each movement. This study will help the industrial engineer determine the best way to complete a specific task and then to suggest modifications or redesigns to accommodate the needs of the user or worker. The results of time–motion studies are used to establish standard times within which tasks should be completed. Unfortunately, these standard work times came at a physiological cost to the workers.

What is a motion study?

The Gilbreths developed a language and a method to understand work motions. This included using film to record the amount of time it took to complete tasks and the motions required to complete the tasks. They looked for areas of improvement, addressed physiological challenges to completing work, and identified the best way to perform the given tasks.

What is a time–motion study?

Time–motion studies are a combination of the contributions of Frederick Taylor and Frank and Lillian Gilbreth. They are used to improve productivity as the motions and activities of different jobs are analyzed. These analyses are used by industries and organizations to develop work standards and set the pace

What types of problems have time-motion studies helped solve?

DID YOU KNOW?

Time–motion studies can be used in health care to improve how healthcare workers use their work hours in healthcare facilities. From these studies, an industrial engineer can develop productivity standards and break down each task into steps, determine the best sequence of motions, analyze ergonomic data to improve working conditions, and measure the precise amount of time needed to complete specific motions and tasks. Another problem that time–motion studies can solve is if a manufacturer is deciding if they want to or how to automate manufacturing tasks.

at which work should be done. It was used as a management technique that at times led to what has been reported as improved productivity and overall work systems that later developed into a field called methods engineering.

The Gilbreths pioneered the idea of using film to document workers and their work. They identified the fundamental elements of work, which they called therbligs (plural of Gilbreth spelled backward). They used a camera and a device used to record the time that the film was exposed (recording the observed activity). This camera was calibrated to capture fractions of minutes of time, which could be associated with their smallest motions.

How do I complete a time–motion study?

An engineer will report time–motion studies until they believe that they have refined the process of completing the specific task under study. Here are the steps to complete a time–motion study: Watch the process until you feel comfortable with it; list the steps needed to perform the tasks; discuss them with the worker (interview them); measure each step with a stop watch as the worker performs the tasks; repeat the entire process at least 10 times; and compute the mean and standard deviation of each step and of the entire task. Keep in mind that the worker may get distraction, need to be responsive to sudden changes, or may need time to learn the equipment, process or procedure.

QUALITY CONTROL

What is a manufacturing process?

A manufacturing process includes the steps an industry uses to transform materials into a product. It is how a company develops products for its cus-

tomers. Products may be produced in batches of various sizes, so it is important for the company to use the manufacturing process that best fits their needs. The process is different based on the materials being used and products manufactured. Example processes include production line, continuous flow, custom manufacturing, fixed-position manufacturing, and discrete manufacturing. Each manufacturing process has a unique layout.

What is lean manufacturing?

Lean manufacturing emerged from Japanese manufacturing as they faced financial, technological, and labor relations challenges. The goal of lean manufacturing is to eliminate waste by determining what practices are value-added (the customer is willing to pay for this work), non-value-added (the customer does not pay for this work but it is essential for manufacturing or quality), or waste (the customer does not pay for it and it is unnecessary for manufacturing or quality).

Who were the main contributors to lean manufacturing?

Some of the most prominent contributors to lean manufacturing were Frederick Taylor, Henry Ford, Sakichi Toyoda, Kiichiro Toyoda, Taiichi Ohno, and Shigeo Shingo. Before lean manufacturing was invented, Henry Ford in-

Called the "Father of the Japanese Industrial Revolution," Sakichi Toyoda founded the Toyoda Loom Works and invented the automatic power loom.

troduced standardization, reduction of waste, and just-in-time manufacturing. Sakichi (Toyoda Loom Works) developed a weaving loom that would automatically stop when the yarn breaks. This is the concept of automation. Kiichiro Toyoda developed the Toyota Production System (TPS), which used lean principles and emerged from the automotive division of Toyoda Loom Works that was founded by his father, Sakichi.

What is the Toyota Production System?

Sakichi Toyoda (1867–1930) is known as the "Father of the Japanese Industrial Revolution" and the king of Japanese inventions. (His son, Kiichiro Toyoda [1894–1952], founded the Toyota car company.) But the influence of his work goes beyond the boundaries of Japan. His automatic stopping loom opened a door for applications in many manufacturing industries. Sakichi has a persistent commitment to continuous improvement, or *kaizen* in Japanese. TPS builds upon this commitment and has a goal of providing customers with their ordered vehicles in the quickest and most efficient way while ensuring product quality to the customer's satisfaction. TPS is built upon a concept called jidoka (the machine must come to a safe stop when a defect in the process is discovered) and a concept called just-in-time (JIT; the process requires and produces only what is needed).

What does a manufacturing process look like?

Manufacturing processes produce a vast majority of the products that we use and see on a daily basis. A manufacturing process changes the form and function of a material to meet a need. This process starts from gathering the raw materials and continues all the way to delivering the finished product. The raw material must be transformed in some way. More materials might be added to it or taken away, or it may be the same material given a new form through shaping (i.e., bending or forging), machining (i.e., drilling, boring, sawing, or grinding), metal forming (i.e., using force to cause permanent change or stretch forming), joining (i.e., welding, soldering, or screwing), finishing (i.e., grinding or polishing), or changing material properties (i.e., flame hardening). The manufacturing process can include any of these activities and more.

What are the most common steps in a manufacturing process?

The manufacturing process spans from product ideation to actual development. This can best be described by the following steps:

THE MANUFACTURING PROCESS

Step	Description
Product ideation /conception	The designer considers what the product will be, how will it be used, and who will use it.
Research	This step requires an in-depth study of the need for the product and what other products, if any, exist that may be similar to the intended product.
Product-design development	This is where the initial design ideas of the product are explored, taking into consideration the functionality, reliability, cost, complexity of manufacturing, usability, and availability of materials needed to make the product.
Research and development often final design	Here, the design is edited and finalized to be sure that the highest quality product is created.
Computer-aided processing	This includes the use of computer software to first create a three-dimensional rendering of the product as well as the manufacturing of the actual physical product.
Prototype testing	The product is tested using nondestructive and sometimes destructive testing to identify any flaws or failures that may reduce the usability of the product.
Manufacturing	Having completed testing of the prototype and fixing any issues that arise, the actual product is manufactured.
Assembly	If the product has multiple parts, this is where these parts are assembled into the final product.
Feedback and testing	The final product is tested in a fairly extensive manner, and the product is assessed by potential users for feedback.
Product development	Using the feedback from the previous step, the product may be further refined.

How is manufacturing engineering similar to/different from industrial engineering?

Manufacturing engineering and industrial engineering have many similarities, but at their core, they are two different disciplines. Industrial engineering is focused on improving the efficiency of people and processes within a work environment or context. This area of engineering takes into consideration people, scheduling, logistics, etc. Manufacturing engineering, on the other hand, is concerned with the machinery and process of producing something. Manufacturing engineering, therefore, deals with the raw materials, tools, machines, and overall process of making a product.

What are some jobs a manufacturing engineer might do?

Manufacturing engineers are often responsible for the design and operation of manufacturing processes to create a product. No matter what the job title, they are concerned with creating the most efficient, cost-effective, and optimized system they can in order to produce something. To this end, a manufacturing engineer may be responsible for:

- Automating a manufacturing process using relevant computer technologies
- Designing a product manufacturing process so as to minimize cost and maximize product quality
- Determining the most optimal layout of a processing plant
- Designing the tools and/or equipment needed to make a product if no such tools/equipment are available or exist
- Identifying cost-effective ways to handle materials in a manufacturing environment
- Quality management and control, supply chain management, inventory control, and forecasting and cost analysis

What are the different types of manufacturing processes?

Currently, five basic types of manufacturing processes exist, with others that are soon emerging:

DID YOU KNOW?

What is an emergent manufacturing process?

3D printing is known as the sixth manufacturing process. The manufacturing industry has many subsectors based on the products they create. Each subsector has its own manufacturing process. Some examples include food manufacturing, textile product mills, apparel manufacturing, wood product manufacturing, chemical manufacturing, and computer and electronics product manufacturing. Considerations a company makes as it selects a process include customer demand, machining techniques and materials, and the availability of resources.

- Process manufacturing (continuous): This is similar to repetitive manufacturing with the exception that the process produces products from raw materials.
- Process manufacturing (batch): This option is highly dependent on the customer demand; thus, the production process is more diverse than repetitive and shares more similarities with continuous manufacturing.
- Job (production) shop: This process does not use an assembly-line approach; rather, it makes use of production areas. The process typically creates a single item. This is common for products that have low demand. Examples of the job shop process layout are bakeries, law firms, emergency rooms, and automobile repair shops. This is also called the flow shop manufacturing process. In this process, resources are grouped according to similarity.
- Repetitive manufacturing: This process is used when a company is manufacturing the same product 24/7 using an assembly-line approach. The speeds can be adjusted.
- Discrete manufacturing: Discrete manufacturing also uses an assembly-line approach. Unlike repetitive manufacturing, this is a robust approach that allows for different projects of diverse dimensions. Production of materials in a discrete manufacturing process takes more time in order to allow for modifications. In discrete manufacturing, the resources needed to create the product are arranged in sequence with functions/operations (when they are needed).

What is plant design?

Plant design concerns the special arrangement of all the machines, materials, people, and resources in a manufacturing facility (plant). This is also referred

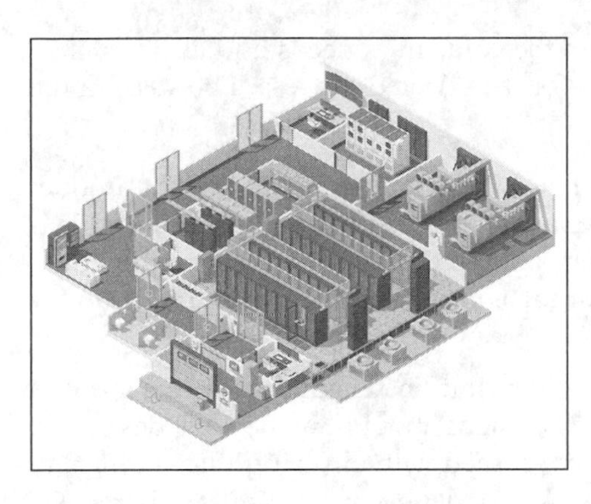

Figure 15: Industrial engineers are involved in facilities planning, which is deciding what is needed in a plant and how best to arrange it.

to as facilities planning. Industrial engineers determine the best arrangement and the appropriate number of machines and workstations needed for the plant to be productive. The primary goal of plant layout and design is to minimize the movement of workers and materials needed to produce the product and to maximize how much of the product can be produced. Other objectives include:

- Minimizing handling of materials
- Maintaining flexibility of operations
- Ensuring optimum utilization of workers, materials, equipment, and available space
- Achieving good work flow and avoiding accumulation of work
- Minimizing delays and bottlenecks in the production system
- Ensuring the safety of workers by minimizing and eliminating the chances of accidents
- Providing for effective supervision and production control
- Minimizing work-in-process inventory
- Providing sufficient and conveniently located service centers
- Flexibility in designs in order to adapt to changing future requirements

Do industrial engineers use a process to design a production facility?

Yes; an industrial engineer works with their marketing management team to ensure that the quantity of the product being produced will meet demand and will not be too expensive. Once the engineers know what the product is and how much to produce, they will then design a production facility that will meet their goals. The engineer then has the responsibility to determine which

manufacturing process will best meet the company's needs and also to determine which equipment is best suited for the selected process. The steps in the process are as follows:

1. Collection of required data: The engineer needs to know what product will be manufactured there, quantity, facility size, space availability, machines needed and their uses, and method of product manufacture. They will ask many questions to people throughout the business to get answers to these questions.
2. Preparation of the floor plan blueprint: The engineer uses the answers from their questions to make decisions about how they will design the plant. They are specifically concerned with ensuring that the facility layout allows for easy and unobstructed movement of materials, people, resources, and products throughout the manufacturing process.
3. Preparation of process chart and flow diagram: The engineer creates a diagram that represents the activities and movement of the materials required to produce the product.
4. Preparation of draft layout: The engineer drafts a layout that includes not only the activities to be completed but also indicates where workers' workstations are, where workers will be, and the material and process flow. They share this layout of supervisors, production workers, and other engineers to get feedback. After they get enough feedback, they make all the changes needed and create the final plant design.
5. Test run: The engineer uses software to simulate the plant design to test for timing, efficiency, material flow, and process. This will give them an idea of how the manufacturing process might work without having to spend and potentially waste money building a whole new facility for testing. Once the engineer is confident in the design, they work with the company to construct their plant layout. In the real-time work environment, they complete test runs to find any additional problems that they did not catch in the earlier phases of the process.

What are facilities?

Facilities are physical locations, such as buildings, that are created to serve a specific purpose where people, materials, products, and equipment work together to meet this desired end.

What types of facilities can engineers design?

- Dance studio

Why is quality control important?

DID YOU KNOW?

Quality control is important because it ensures that a process is in place to inspect the quality of the products produced and control process variations that impact product quality. The goal is to fulfill quality expectations for any product of interest. In some industries, this could be a physical object like a tennis ball or automobile. For service industries, it could be reports, contracts, or designs. Quality control can be subdivided into (a) statistical analysis, (b) probability theory, (c) reliability analysis, and (d) design of experiments (DOE).

- Manufacturing plant
- Grocery store
- Hospital
- Assembly department
- Existing warehouse
- Baggage department of an airport
- Production plant
- Retail store
- Dormitory
- Bank
- Office
- Cinema
- Parking lot
- Any portion of the above

How do engineers measure quality?

During inspection, the engineer will measure, examine, and test specific components of the products. Then, the results of these tests are compared against the requirements to determine whether they are conforming to standard or if a deviation from what is expected has occurred. Basically, the industrial engineer is inspecting the product and asking, "Is this the expected product,

and does it meet specifications?" If it does not, then they can review the manufacturing process, system, materials, and other aspects to determine why the product does not pass inspection.

What is statistical analysis?

Engineers use statistical methods to ensure and improve quality. They collect data reports and compare them to established quality standards. Statistics is a branch of mathematics used to study measurements and data. It applies to quality engineering because of its ability to tabulate, depict, describe, and predict how groups behave. Here are some basic descriptive terms that everyone should know about statistics. These terms help engineers describe the work they are doing.

- Population: A group of items, people, or communities that is being studied
- Sample: A subset of the specific population being studied
- Random samples: Every member of the sample has an equal chance of being selected
- Data: Information, numbers, or measurements that are being collected in a population
- Variables: Characteristics used to distinguish data members from one another
- Constant: Items and characteristics of the members that will never change
- Data set: A collection of data that accords with members and their characteristics

Statistical analysis is necessary for engineers to improve the quality of any project or process.

What are some common statistical approaches?

Some statistical approaches that an industrial engineer might use are these:

- Hypothesis testing: Reject or fail to reject the null hypothesis
- Regression analysis: Develop mathematical equation to describe the relationship between a response variable and independent variable(s)
- Statistical process control: Monitor, control, and improve a process by identifying and removing variation in the process
- Design an analysis of experiments: Use a well-designed and controlled test to evaluate factors that influence a response variable

What software programs do industrial engineers use?

Industrial engineers often use software programs such as Excel, R, STATA, SAS, or SPSS to perform calculations. These software programs can be programmed to follow specific instructions to complete the calculations for a given data set.

What are some common statistical calculations?

- Mean: The average of all the data items. (Example: Twenty students are taking a scored exam. Although they may have different scores, their average exam score tells the teacher what the typical score in the class is.)
- Median: The middle number in a data set. In the same class of 20 students, the scores are ordered from least to greatest in order to create a range. The median is the score that lies at the midpoint of the range.
- Variance: The average of the squared difference of each item from the mean. The difference of each item from the mean is squared because otherwise, negative and positive differences would cancel each other out.
- Standard deviation: How spread apart the items in the data set are. Mathematically, this is the square root of the variance.

- Standard error: The spread of the means of multiple samples from the same population. If the samples are an accurate representation of the population, then the standard error should be close to zero.
- Z-score: How far a specific observation (datum) is above or below the mean of all the data points in that data set.

Who was W. Edwards Deming?

W. Edwards Deming (1900–1993) was an influential, world-renowned contributor to the realm of statistical analysis and quality control. He made major contributions in both the United States and Japan. His expertise was initially developed during his work on the 1940 U.S. Census. He would later apply statistical quality control to other governmental agencies, such as the USDA. His teaching of the application of statistics to address quality issues was well received in Japan. The United States was slower to embrace his teachings.

What is probability theory?

Probability theory is a calculation of the likelihood that a specific event will occur. This likelihood is represented by a ratio. The probability that an event will occur equals the number of ways it can happen divided by the total number of outcomes. Consider a two-side coin with a face on one side and an eagle on the other. The probability that a coin toss will land on the eagle is equal to 1

An electrical engineer and physicist, W. Edwards Deming was an author, professor, statistician, and management consultant who was a renowned expert on statistical analysis and quality control.

(number of ways it can happen) divided by 2 (total number of outcomes): .5. Probability falls between 0 and 1, where 0 means the event cannot occur and 1 means that the event is certain to occur.

What is reliability analysis?

The reliability of a product or process is determined by the probability that the product or process will function as it was designed for a specified period of time. Industrial engineers determine reliability by observing failures in that given time period. An example of a reliability issue was the Firestone tire recall of 1996. Engineers calculate the failure rate for a large number of products tested over time. This helps them determine for how long products are expected to function as designed. Companies often keep records of failure and repair for their products. They use this information to determine how and when to provide support to the user and also when they are redesigning products.

What is design of experiments (DOE)?

Design of experiments is very important to industrial engineering. DOE is a tool that industrial engineers use to understand processes, conditions (variables) that influence the processes, and how to improve the process. IEs must understand how different conditions (variables) affect the quality of manufactured parts. Sometimes, it is difficult to model a complete system at full scale. This is especially true if many variables must be measured or assessed. The four steps of DOE are design the experiment, collect the data, analyze the data using statistics, and develop conclusions and recommendations from what you learn.

ERGONOMICS

What is ergonomics?

Ergonomics is the study of a human's work and achieving optimal work settings for the environment. Human work often requires interaction with machines and equipment. Industrial engineers study ergonomics in order to improve this interaction. This impacts worker performance, worker health, and satisfaction with their job responsibilities. Ergonomics is broken into two areas of study: occupational ergonomics (human factors) and cognitive ergonomics (task analysis). Occupational ergonomics is concerned with human ability and

DID YOU KNOW?

Ergonomics sounds like economics. Are they similar?

Although some say that good ergonomics is good economics, they are certainly not the same thing. Ergonomics is the study of human ability and behavior as they interact with machines. Economics is the study of the production, consumption, and transfer of wealth. While these two words sound similar, they are not the same. However, a relationship does exist between the two areas of study: good design decisions often save companies great amounts of money and resources in the long run.

their interaction with machines, whereas cognitive ergonomics is concerned with human behavior and their interaction with machines.

What are some examples of how good ergonomic decisions made sense economically?

One example of a cognitive ergonomic problem is this: An aircraft throttle system needed modifications because it was too sensitive to pilot inputs. This was a big concern, as it made it difficult to accurately control the airplane speed. Once this sensitivity was observed, industrial engineers proposed modifications to the throttle system, which saved the company from doing a very costly redesign of the entire system.

What are some examples of ergonomics in the healthcare industry?

A well-designed work procedure can save a company money in many ways, from paying medical expenses to saving the employed worker from long-term musculoskeletal disorders, doctors' visits, and eventual worker's disability. A telecommunication company successfully overcame its occupational ergonomics challenge by making various occupational ergonomic improvements and achieved a $1.48 million reduction in workers' compensation costs and reduced the number of lost workdays due to work-related injuries. These two examples show that good ergonomics is good economics, but they are not the same!

What are some examples of ergonomic problems?

Some examples of tasks that occupational ergonomics is concerned with include lifting, turning, stretching, bending, reaching, pushing, pulling, reaching overhead, and working in awkward body positions. They may also study the environmental effects on human workers and their ability to complete tasks. Those might include the temperature of a workspace, humidity, or the vibrations of the room or the tool.

Why do industrial engineers study the long-term impacts of working conditions?

Some of these tasks and the environment rarely lead to fatalities. Industrial engineers concerned with occupational ergonomics are more concerned with the long-term effects on the body. Musculoskeletal disorders are one of the more common work-related injuries sustained by workers. In the long term, this leads to persistent pain, can reduce workplace performance, and could ultimately lead to disability. Industrial engineers are trained to study human workers, how they use their workplace tools, and the impact that the tools have on their bodies in order to determine where injury might occur and develop solutions in order to prevent them.

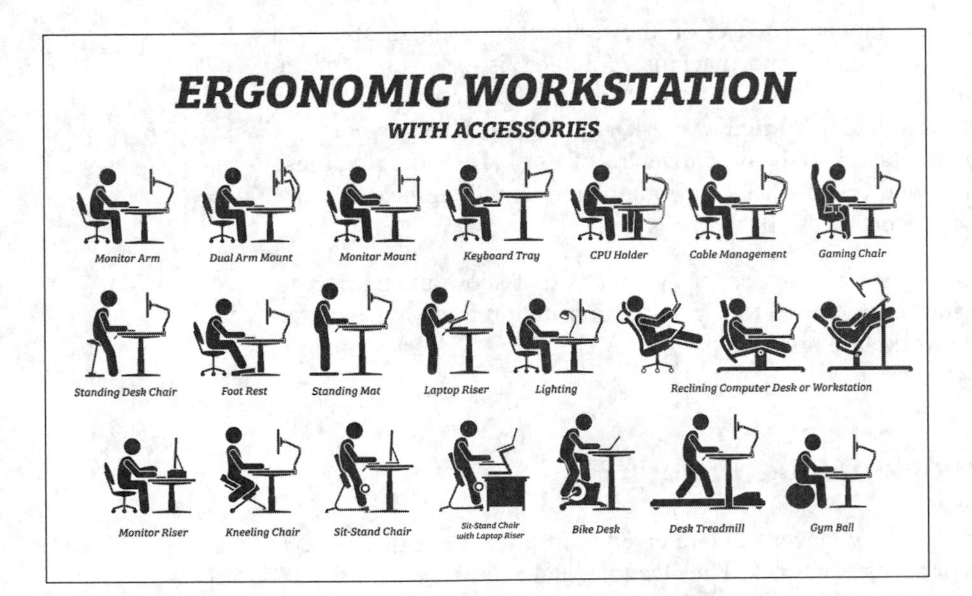

Because so much work is done at a desk these days, when people say "ergonomics," the first thing that might pop into their brains is how workstations are set up (as in these examples). Ergonomics involves other factors, too, such as workplace temperature, lighting, and humidity control.

DID YOU KNOW?

What is cognitive ergonomics?

Human behavior is important to study because it helps the industrial engineer understand if the human knows how the machine works, how to interface with the machine, and how to design machines that support the work that humans are required to do to complete a task.

What are some common ergonomic injuries?

- Carpal tunnel syndrome
- Tendinitis
- Rotator cuff injuries (affects the shoulder)
- Epicondylitis (affects the elbow)
- Trigger finger
- Muscle strains and lower back injuries

What is occupational ergonomics?

Occupational ergonomics focuses on the study of the human ability to design work spaces, tools, jobs, and machines. The goal is to ensure that the work environment is safe. Ergonomists evaluate jobs by looking at three main stressors: the force required to complete a task, any awkward worker postures adopted to complete a task, and the repetitiveness of a task. By observing these three stressors in how a worker does their job, the industrial engineer can determine the impact of the job on the worker's body.

Some companies are creating special positions that allow an engineer to design and 3D print new tools and tool modifications that reduce stress on the body and improve workplace productivity.

What is a work-related injury?

A work-related injury may also be referred to as a work accident, occupational accident, or accident at work. The Occupational Safety and Health Ad-

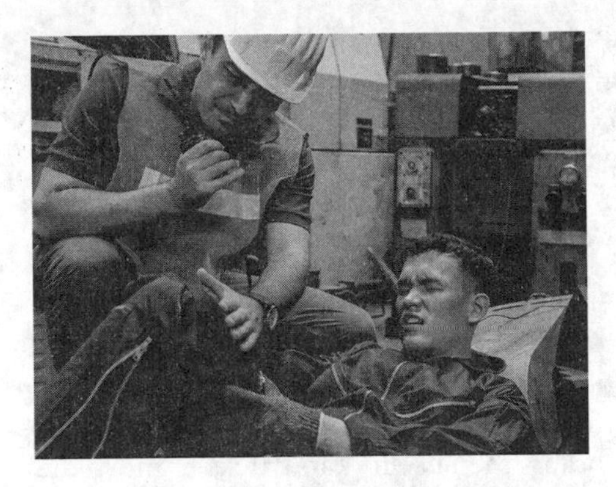

Work-related injuries are an ever-present danger when employees are operating machinery. Government standards set up by OSHA are designed to try to make such occurrences less common.

ministration is a division in the U.S. Department of Labor that is concerned with ensuring safe and healthful working environments, enforcing standards, and providing resources for industries to meet and exceed standards. According to the OSHA standards, a work-related injury or illness is one in which the initial event or exposure occurred in the work environment. This includes both the physical work location and the tools used by the employee to complete their work responsibilities.

MATERIAL HANDLING

What is material handling?

Material handling is providing the right materials at the right place, at the right time, and for the right costs. Attention is given to how the manufacturing plant is designed to ease material handling between workstations. Manufacturing products requires access to many different types of materials and, depending on the process, at specific times. Material must be moved, stored, controlled, and protected throughout all points of the process. The material handling field is primarily concerned with how to achieve movement, storage, control, and protection of materials.

Why is material handling important? What happens if material is delivered out of sequence?

Material delivered in the wrong sequence will result in lost time and resources. Not having material at the right time means that production will be

delayed. One of the primary goals of implementing a material handling plan is to reduce the unit cost of production. Basically, an industrial engineer will design a plan that includes determining how material will move through the system so that the total cost to produce any product is as low as possible without sacrificing product quality.

What is palletizing?

Palletizing means using pallets for storing and transporting goods. Materials can be stacked on pallets, and workers then use forklift trucks to move the pallets from storage to wherever the materials are needed.

Where are materials stored?

Warehouses are large building structures where materials needed for production are stored. In some cases, some materials, such as those to be used in chemical and biochemical manufacturing processes, must be stored under climate-controlled conditions. In such instances, climate-controlled warehouses are used. These types of warehouses can range from humidity-controlled environments to freezers.

What are material handling systems?

Material handling systems refer to the types of equipment used to move, store, and control materials through a variety of processes at warehouses and

Wooden pallets like these are used as a base to stack goods for shipping.

Conveyor systems can move heavy materials quickly and efficiently through a facility.

other similar facilities. In industrial engineering, this is often an automated process and includes the use of engineering systems such as automated storage and retrieval systems (AS/RS), automated guided vehicles (AGVs), robotic delivery systems, and conveyor systems.

When designing material handling systems, industrial engineers should consider (1) performance objectives, (2) standardization, (3) ergonomics, (4) space, and (5) automation.

What are conveyor systems?

Conveyor systems are very helpful when very heavy materials must be moved from one point to another. Conveyors are very common in a manufacturing facility because of their ability to transport materials in a quick and efficient manner. Conveyors can also be found in grocery stores, airports, and other transportation facilities. Conveyor systems come in many types; some of the most commonly used in engineering environments are pneumatic, vibrating, flexible, spiral, vertical, and motorized drive roller conveyor systems.

METHODS AND PROCESS ENGINEERING

What is process engineering?

Process engineering focuses on determining the process and methods to complete work. An existing work procedure consists of five stages to analyze

the status of the work methods selected: project selection, data acquisition, analysis, development, and implementation. Whereas other processes described are more focused on resources, methods engineers determine where in the process humans engage to get the task done and how they will complete tasks. Basically, methods engineering is concerned with activity within a workplace.

Process engineering is also known as operations analysis, work design, and simplification. The overall goal of this type of engineering is to minimize the cost while increasing the reliability and productivity of an engineering process. Methods engineering is often concerned with selecting a project, acquiring and presenting data, analyzing said data, identifying the ideal mode of development, and presenting and implementing the selected method.

What is the difference between a process engineer and a manufacturing/production engineer?

A process engineer tends to work within a system where the core focus is on continuous manufacturing. In their jobs, they work with raw materials destined to become a final product, often through the application of chemical processes.

A production or manufacturing engineer works with machinery and other equipment that are individual parts of a larger product. These smaller parts, often part of an assembly line, are then assembled to form the final product. This is known as discrete manufacturing.

What does a process engineer do?

As the title suggests, a process engineer designs, implements, and optimizes engineering processes. They often work in various engineering contexts but most commonly in chemical or biochemical facilities, food processing, pharmaceutical, and other agriculture-related industries.

What is a methods engineer?

A methods engineer evaluates manufacturing processes to determine where and how humans can be integrated into the process in an efficient and effective manner. For example, in the process of converting some raw materials into a finished product, a methods engineer would determine where human skill and expertise would be best suited.

What is operations analysis?

Operations analysis seeks to understand customer results, strategic and business results, financial performance, and innovation. This method analyzes the current process against established or desired standards with respect to costs, schedule, and performance standards.

What are the steps in method engineering?

These steps allow method engineers to improve organizations and processes and continually improve and fine-tune the organization's approaches to problem solving.

1. Define the problem and objectives
2. Collect and analyze data
3. Formulate alternatives
4. Evaluate alternatives and select the best one
5. Implement the best method
6. Assess the method's engineering project

SIMULATION ANALYSIS

What are simulations?

Simulations can be simplified as imitations of real-world scenarios. This can be done through a physical simulation in which physical objects are used to model the scenario. An interaction simulation can be used to help workers or users of the system; consider flight simulators or driving simulators.

What is simulation analysis?

Simulation analysis is a technique that industrial engineers use to examine a system and evaluate how it will behave if operation conditions change. It can also be done on a new or proposed system. Simulation analysis can also take the form of computer simulation, which is a common tool that engineers used to imitate and analyze many different types of scenarios. This could be as simple as designing a tool and analyzing how much stress it can handle. Some computer simulation software programs include ProEngineer, SolidWorks, and ANSYS. These are called computer-aided design programs.

What tools do industrial engineers use to simulate real-world processes and procedures?

- ARENASolidworks for simulation
- Tableau for data visualization
- Factory-level production management system
- Computer-integrated manufacturing (CIM)
- Computer-aided design (CAD)
- Computer-aided engineering (CAE)
- Computer-aided design and drafting (CADD)
- Computer-aided process planning (CAPP)
- Computer-aided tool design (CATD)
- Computer-aided manufacturing (CAM)
- Computer-aided numerical control (NC) part programming

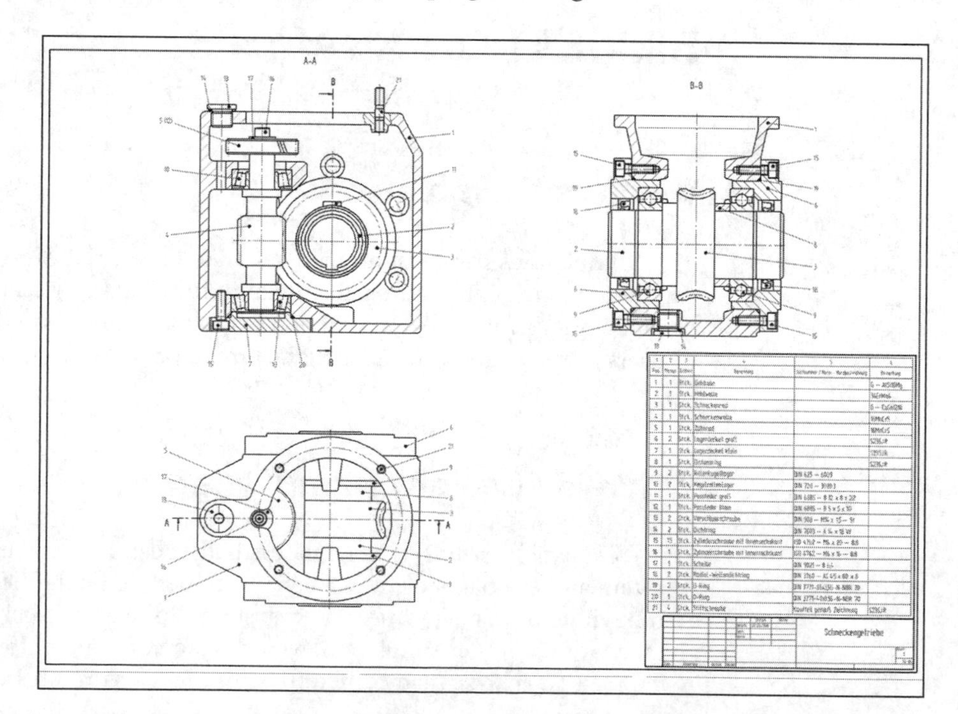

An engineering drawing executed with a CAD software called Solid Edge. This one is in 2D, but 3D drawings are also possible.

DID YOU KNOW?

What is human-technology interaction?

In many workplaces, humans work with machines, technology, and tools to complete their job responsibilities. Industrial engineers are responsible for observing and understanding humans' interaction with the technology required for their jobs or leisure and ensuring that the interaction doesn't lead to long-term disorders or performance ineffectiveness. For example, a pilot interacts with a throttle system to control speed, or a homeowner uses a power drill to complete household tasks. Both examples show an interaction between a human and technology. Interactions like these and many others must be well understood and, if found to need modification, be improved to reduce negative impacts.

- Computer-aided scheduling (CASCH)
- Computer-aided material requirement planning (MRP)
- Flexible manufacturing system (FMS)
- Group technology (GT)
- Computer-aided testing (CAT)

OPERATIONS RESEARCH

What is operations research?

Operations research (OR) is an industrial engineering topic that comes from its history of military operations during World War II. Since World War II, the area of study has advanced to address issues in the context of human–technology interaction and decision making. Industrial engineers with a focus on operations research use mathematical modeling and other scientific methods to study complex systems, predict outcomes, and mitigate risks. These problems are complex because they require engineers to address a large number of variables, resources, and constraints.

What is resource allocation?

Resource allocation is the process of assigning resources and materials to complete tasks. Resource allocation is important to industrial engineering be-

DID YOU KNOW?

What is a mathematical model?

When a problem or system is simple, it may be much easier to use trial and error to find the solution. When problems and systems are complex and have many variables, constraints, and resources, engineers develop mathematical models of the systems. Engineers apply mathematical methods and tools to help them model the systems and explore the impact that changing variables have on the systems' functions and operations.

Mathematical modeling is just one tool that operations research experts use to solve complex problems. Other tools include linear programming, game theory, queuing theory, and decision analysis.

cause improper resource allocation can mean increased costs, delays in production, missed deadlines, and overall inefficiency of the production process.

What are typical problems that OR can help solve?

Some of the typical problems that operations research can solve include:

- Scheduling and logistics for airplane travel (air traffic control, passenger ticketing)
- Package delivery (product stock and retrieval)
- Emergency services (911, catastrophic event support)
- Theme park experiences (ticket prices, ride queue length, ride time)

What is linear programming?

Linear programming is a mathematical tool used by industrial engineers to address complex systems. Industrial engineers want to determine the best outcome given specific constraints. For example, it may be valuable to know how to get maximum profit from or the lowest costs for the products the facility produces. This technique means that the industrial engineers know exactly what the inputs are and want to determine the best output given those inputs. For example, every production facility has limitations on workers (laborers) and materials. These are inputs for the linear program. The industrial engineer has to determine based on those inputs what is the best production level in order to achieve the maximum output and lowest cost to produce.

Is linear programming similar to writing code in a programming language?

A linear program is not the same as writing a computer code, but they are similar because they both use algorithms. In computer programming, algorithms are used to construct a code. In linear programming, algorithms are used to solve linear programming. George Dantzig is famous for developing an algorithm to solve linear programming problems. His solution helps industrial engineers address issues such as where to allocate resources, materials, and time.

What is the future of industrial engineering?

- Advancement in technology: The automation of tasks and manufacturing and the ways that new information changes industries and our approaches will force industrial engineering to shape the role and function of industrial engineers.
- Health care: Due to improvements in the access of medical information and knowledge, doctors are no longer the sole source of information but must learn how to help patients make sense of information, correct misinformation, and make healthcare decisions.
- Manufacturing: In these industries, the Six Sigma process improvement processes are accessible online for many people to acquire. Industrial engineers still provide much needed expertise not only with the process but also mastery of the tools needed to advance these industries.
- Big data: Advancements made to manufacturing facilities and in computer science allow for more data to be generated and stored related to production processes, machines, and material flow. Industrial engineers are developing ways to make use of the vast amounts of data being generated all over the world in various contexts to help people make data-driven decisions. This crosses the boundaries of data science and machine learning and can influence manufacturing, health care, medicine, and politics.
- Mobility: This is becoming a top priority for high-level industrial engineers in the industry. Mobility has to do with goods, services, and people interacting and moving around more freely. One role of the industrial engineer is to determine the process by which materials, people, and serv-

Part of intelligent manufacturing systems is digital manufacturing. Using software programs like Robcad, for example, engineers can simulate a production line with robots and run animations to test how well the system might work.

ices are moved and produced. Increased interaction and mobility have huge implications and increased demand for the industrial engineer.

- Intelligent manufacturing systems (IMS): As technology continues to advance and influence every aspect of our lives, manufacturing systems must adapt as well. Intelligent manufacturing systems will integrate the skills and knowledge of humans with machines, equipment, and processes to ensure the best possible manufacturing outcome. These systems will use information to optimize resources, minimize waste, and add value to the organization in which they operate. Intelligent manufacturing systems include digital manufacturing, digital-networked manufacturing, and new-generation intelligent manufacturing.
- Smart systems: Similar to IMS, smart systems will incorporate the use of sensing technology to make decisions about the manufacturing process so as to maintain the highest levels of efficiency, control, and actuation. Smart systems are capable of identifying challenges to the manufacturing process such as limited resources, changes in the production timeline, and how other social, environmental, and economic challenges might impact the process.

Bioengineering and Biomedical Engineering

What is biomedical engineering?

Biomedical engineering is a multidisciplinary field that brings biology and engineering together to apply engineering concepts and materials to medicine and health care. It applies engineering ideas and design concepts in medicine and biology with the goal of providing health care. Biomedical engineers have created a variety of devices to address medical needs and improve and save lives. This includes artificial organs, surgical robots, advanced prosthetics, systems to monitor vital signs, new pharmaceutical drugs, imaging methods (such as ultrasounds and X-rays), kidney dialysis, radiation, and physical therapy.

What is bioengineering?

Bioengineering consolidates science, engineering, and innovation to make devices, materials, and procedures for improving human well-being. It centers around systems and solutions to protect the environment and to improve controlling measures for ecological dangers. Hence, bioengineering materials and consolidated systems are needed to diminish the negative impact humans have had on Earth.

What are some examples of bioengineering?

Examples of bioengineering research include bacteria engineered to produce chemicals, new medical devices such as imaging technology equipment, portable and rapid disease diagnostic devices, prosthetics, biocompatible materials such as prostheses or therapeutic biologicals, biopharmaceuticals, tissue-engineered organs, and regenerative tissue growth processes.

Are biomedical engineering and bioengineering the same thing?

Bioengineering is a vast field that comprises biomedical engineering, medical engineering, and biochemical engineering. In contrast to biomedical engineers, bioengineers center around making new items, such as pharmaceutical products, food supplements, preservatives, bionanotechnology, and biomass-based energy, using fundamental ideas and procedures in the biological sciences. Biomedical engineering, on the other hand, uses major standards and fundamental principles of natural sciences, medical sciences, and engineering to improve human well-being.

Thus, bioengineering is a worldwide term that includes biomedical engineering and is connected to all life sciences and medicine, while biomedical engineering focuses on medical and healthcare services.

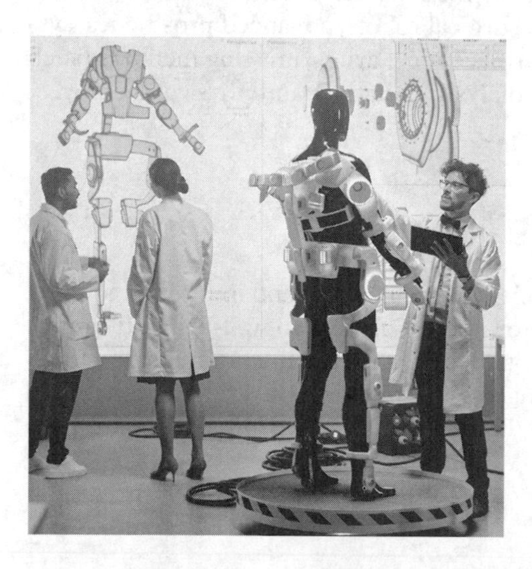

Biomedical engineers work in a wide variety of specialties, including bionics. For example, advances are being made in creating exoskeletons that can make it possible for paraplegics to walk again.

What is the history of biomedical engineering?

Biomedical engineering has developed throughout the years in light of progression in science and innovation. The earliest example is a wood and leather prosthetic toe found on a 3,000-year-old Egyptian mummy. Before that, even the first person to design a splint for a broken bone could be considered to have been an early biomedical engineer.

In 1962, Case Western Reserve University included an undergraduate elective sequence in biomedical engineering. Later, the Department of Biomedical Engineering at Case Western Reserve University was officially approved by the Board of Trustees on May 9, 1968.

For many years, biomedical engineers have made progressively more efficient devices to analyze and treat diseases and help the handicapped and wounded with their disabilities and injuries. For instance, the advancement of hearing aids, devices to alleviate hearing disabilities with the help of sound amplification, is an example of biomedical engineering. Some remarkable figures throughout the existence of biomedical designing and their commitments include Wilhelm Roentgen for X-rays, Rene Laënnec for the stethoscope, John Charnley for artificial hip replacements, and Charles Hufnagel for artificial heart valves.

In recent years, biomedical engineering has emerged as a separate study in contrast to numerous other engineering fields. Such advancement is normal

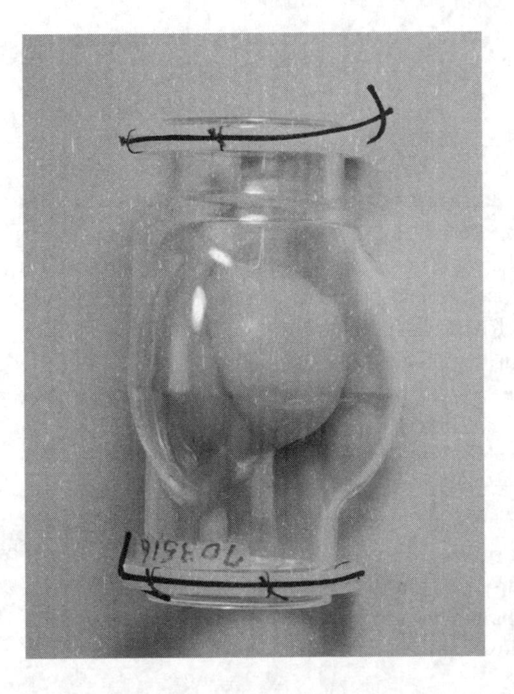

Invented by American surgeon Charles Hufnagel, the Hufnagel artificial heart valve (shown here on display at the National Museum of health and Medicine) replaced the aortic valve. It was first used in the 1950s.

as a field transitions from being an interdisciplinary specialization among already established fields to being viewed as a different field in itself. A significant part of the work in biomedical designing is comprised of research and development.

What are the branches of biomedical engineering?

- Tissue
- Genetic
- Neural
- Pharmaceutical
- Medical devices
- Medical imaging
- Bionics
- Clinical

FUNDAMENTAL CONCEPTS WITHIN BIOMEDICAL ENGINEERING

What is biology?

The structure and processes of the cell, as well as the properties of specific cells in the body, force biomedical engineers to be creative in helping doctors treat conditions. The body has many types of cells, and each of those cell types has a shape and function. Blood cells either carry oxygen (red) or fight diseases (white). Nerve cells carry signals. Muscle cells expand or contract to create movement. Bone cells provide structure for skeletons. Cells make up tissues, tissues make up organs, and organs make up body systems. The cell's abilities affect the abilities and response to injury or disease of the tissues, organs, and systems that they make up. For example, cells reproduce through meiosis or mitosis. Meiosis is the process by which sperm and egg cells are generated. Other cells reproduce identical cells through mitosis. However, not all cells reproduce. That means that certain cells are kept for a person's whole life, and when they die, they are not reproduced or replenished. Certain organs and body systems that contain these cells cannot repair themselves, and diseases could lead to the need for transplants. An example of this is nerve cells; this is why stroke victims go through rehabilitation or surgery but often lose or have diminished function after a stroke.

DID YOU KNOW?

What is fluid dynamics?

Fluid dynamics describes and explains how liquids and gases move. The pumping of the blood from the heart throughout the body using vessels and arteries is in a closed system. Fluids have properties that impact how they move, and water, blood, chemical elements, and medicines move within cells, tissues, organs, and systems. We breathe gases in, our body processes them so that we can use oxygen, and then, we exhale them.

What is electricity?

Electrical impulses that occur in the body on the cellular and tissue levels help with diagnostics, measurement, and treatment of short- and long-term illnesses. Your heart has an electrical impulse that creates a rhythmic wave, which can be read on an EKG. The shape of the wave can show what parts of the heart are not working properly. That is why when the rhythm of a heartbeat is incorrect, a defibrillator can be used to shock the heart in an attempt to get it back to the correct rhythm. The brain produces waves that can be read on an EEG, and the wave activity will show how much and which parts of the brain are working properly when someone has become ill or suffered an injury. Hearing treatments include hearing aids that take sound waves and amplify or compress them based on frequency filters or duplicate the signals that would travel along the auditory nerve.

What are materials?

The compatibility of materials and chemicals helps determine what can be used on the body, in the body, and left inside the body in the form of dissolvable sutures, implants, etc. Nurses and doctors ask patients if they are allergic to latex because sterile gloves often contain latex. Often, bones are repaired with screws that must not rust inside the body, and hips and knees may be replaced with metal parts. Artificial organs are often made from natural or synthetic materials, ceramics, metals, or polymers.

What is physics?

Physical forces act on the body, and body movement or lack of movement is based on balanced forces. The main force that acts on the body is a gravitational force as the spinal cord, bones, and muscles work together to support the various masses (head, arms, and trunk). The spinal cord is S-shaped to increase stability, and discs between the vertebrae cushion applied forces. The body has and is impacted by energy transfers such as potential and kinetic energy. Some other forces that act on the body are pressure, torque, velocity, friction, and acceleration. Many forces of the muscles involve levers that are first-, second-, or third-class lever systems. Injuries often happen when sudden deceleration or dissipation of momentum occurs over a short period of time, which causes pressure or stress on a particular body part. Physics is also used in ultrasound, radiography, X-rays, radiography, and magnetic resonance imaging (MRI) techniques for visualizing the body's soft and hard tissues.

What is mechanics?

The body produces motion while following the rules of mechanics. Muscle forces move the body by contraction or extension of muscles.

What is chemistry?

How chemicals work together and with the properties of the chemicals in the body impact how medicines are made and delivered to the body. Temperature and pressure have effects on characteristics of elements that exist and enter the body. Atoms make up molecules, and molecules make up compounds that make up medicines. The age and size of a person impact the dosage of medication. Understanding how medicines will dissolve in the blood and be carried throughout the body help determine the timing of a dosage and understanding which medicines will counteract or cause bad side effects if taken together helps to prevent negative reactions and patient death. The size and shape of molecules also determine which parts of the body they can enter. For example, certain brain conditions are hard to treat with oral medicines that move through the bloodstream because the blood–brain barrier places size boundaries on the chemicals that can enter the brain. Some chemicals have addictive properties, but they have to be used as medicines, so they must be given in controlled manners, and patients have to be tapered off of them for safety.

What is anatomy?

Biomedical engineers have to understand how the body works so that they can develop solutions that can properly address issues of the body. Anatomy is

DID YOU KNOW?

What is rehabilitation engineering?

James Reswick (1922–2013), a pioneer in this field, said that "rehabilitation engineering is the application of science and technology to ameliorate the handicaps of individuals with disabilities." The objective of rehabilitation engineering is to use technology to help an individual move from a current physical state to an improved state. This can also help to reduce the negative impact of poorly designed systems for people with disabilities. The need for this field was most apparent with the return of wounded World War II soldiers. Scientists and engineers collaborated to develop technologies that could aid veterans. These technologies included such items as artificial limbs, electronic travel guides, and rugged wheelchairs.

the study of the internal and external structures of the body, which helps engineers learn its many functions.

What is physiology?

Physiology is the study of how the body functions at the cellular, tissue, and organ system levels. The study of physiology helps the biomedical engineer know how the various functions interact, communicate, and work together to keep human (or other organisms) alive.

What principles do rehabilitation engineers use to ensure the design the correct assistive technology?

- The users' goals, needs, and tasks must be clearly defined and confirmed early in the intervention process
- Involve rehabilitation professionals with a wide variety of skills, knowledge, and experience on the team
- Perform thorough research to understand the users' preferences, abilities, limitations, and day-to-day lives.
- Explore all existing similar and available technologies so that all useful design concepts are considered.
- Make sure to get user insights and feedback thoughout the entire process.

- Ensure that the assistive technology is installed in the optimal location and fashion for the user.
- Develop and provide training for the user and those that provide day-to-day care. They should be aware of all the technology can do, along with its limitations.
- Within six months of delivery and implementation, follow up with the user and make any changes necessary to better meet their needs.

What are some examples of problems solved by biomedical engineers?

- Synthetic skin for burn victims, developed by surgeon Dr. John F. Burke and by Dr. Ioannis V. Yannas, a professor of polymers and fibers
- Improvements in prenatal surgery that is performed on babies before they are born
- The development of artificial organs by growing tissue transplants for the heart, spinal cord, and brain from the patient's fatty cells

What topics are covered in biomedical engineering courses?

Biomedical engineering focuses on engineering and biological sciences. The program includes classroom-based courses and labs in subjects such as

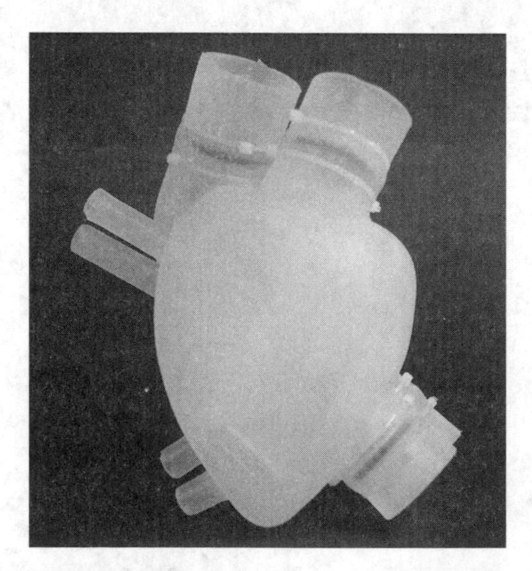

Advances are constantly being made in artificial organs such as this artificial heart currently in development at the ETH Zurich functional material laboratory in Switzerland.

fluid and solid mechanics, computer programming, circuit design, and biomaterials. It also includes substantial training in engineering design.

What software programs do biomedical engineers use?

Biomedical engineers use image processing, data analysis, machine learning software, CAD, Solidworks, and MathCAD. Programming software and skills include MATLAB, Python, C, C++, and LabVIEW.

CAREERS WITH BIOMEDICAL ENGINEERING DEGREES

What are some career choices in biomedical engineering?

Career choices in biomedical engineering are usually driven by the interests of an individual. The huge depth of the field enables biomedical engineers to develop and practice specialties in a particular area that interests them. The area includes subfields in neuromodulation, biomaterials, devices, stem cell engineering, and orthopedic repair.

With a biomedical engineering degree, some work as biomedical engineers while others attend medical schools and become doctors with specializations in various techniques, such as using electric impulses in new ways to get muscles moving again. People interested in research become research scientists. Some earn law degrees and become lawyers. People interested in business administration move into managerial positions.

What is the role of a software engineer in biomedicine?

The role of software engineers is significant in numerous aspects of biomedical engineering. Since the functionality of medical devices is mostly based on software, creating and maintaining these software programs is a significant task for biomedical engineers to tackle. Software engineers develop algorithms for data analysis and biological system modeling. Also, they create systems that guide the clinician in medical records, patient diagnosis, patient monitoring, and clinical decision making.

What do biomedical engineers do?

Biomedical engineers mostly work on designing biomedical equipment, such as artificial internal organs, and on the installment, maintenance, repair, and replacement of machine parts used for diagnosing medical issues. They also give training in the proper use of biomedical equipment. Thus, they develop solutions to medical issues through the integration of science, medicine, and engineering. Biomedical engineers may be called upon in a wide range of capacities: to design instruments, devices, and software; to model the mechanics of the body; to research materials acceptable to the body; or to conduct the research needed to solve clinical problems.

Can the field of materials science be considered a sub-branch of biomedical engineering?

Yes, the field of materials science can be considered a sub-branch of biomedical engineering. Biomedical engineers have built up various life-enhancing and life-saving technologies, including prosthetics such as dentures and artificial limb replacements. The field of materials science deals with various biocompatible materials that are utilized in human bone implants. This field works on developing new metals/materials that are compatible with human tissues and bones.

How is mechanical engineering related to biomedical engineering?

Mechanical engineering has a branch of materials science that deals with various materials and their composition, properties, and behavior. Some metals are used in biomedical applications, one of which is Ti-6Al-4V. It is most commonly used in human bone replacements.

What are some examples of specialty areas in the field of biomedical engineering?

- Bioinstrumentation: The utilization of electronics, software engineering, and measurement principles to build instruments used in the diagnosis and treatment of therapeutic issues.

- Biomaterials: The study of naturally occurring or laboratory-designed materials that are utilized in medical devices or as implantation materials.
- Biomechanics: The study of mechanics—for example, thermodynamics—to take care of therapeutic issues.
- Clinical engineering: The application of medical innovations in order to advance healthcare delivery.
- Rehabilitation engineering: The study of developing devices that help people recovering from or adapting to physical and psychological impairments.
- Systems physiology: The study and understanding of how frameworks inside living life forms, from microbes to humans, function and react to changes in their living condition; it is based on engineering instruments.

What are some major challenges of the biomedical and clinical engineering fields?

Some major biomedical and clinical engineering challenges are as follows:

- Coordinating equipment service schedules: Keeping track of the service needs for each piece of machinery and equipment may be a difficult task, especially if a biomedical engineer is responsible for multiple facilities' equipment. Also, manufacturers' service recommendations are

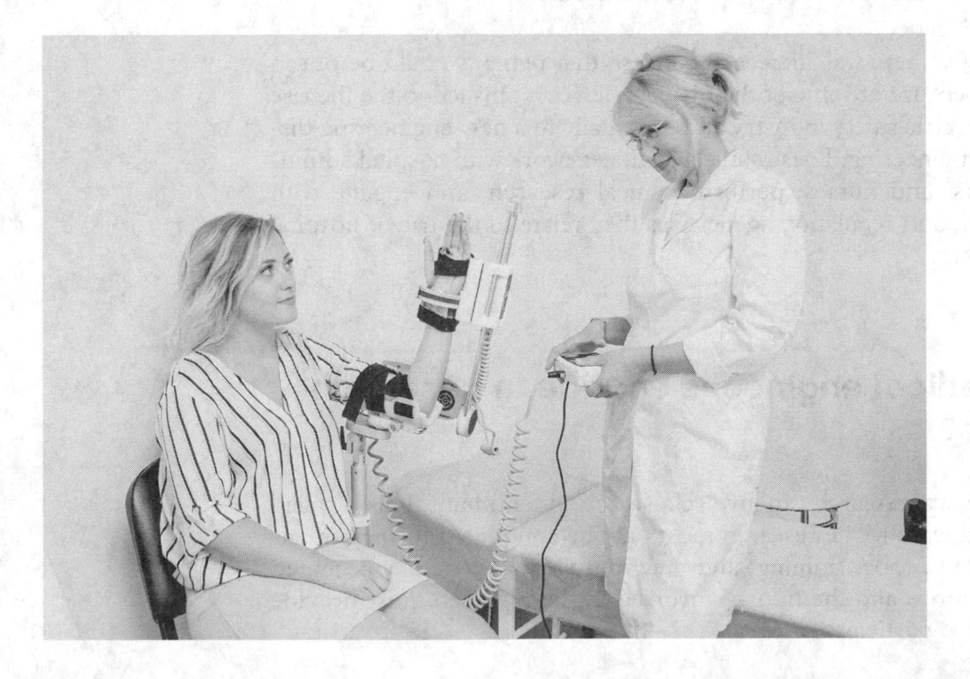

Rehabilitation therapy is one specialized area of biomedical engineering. Devices such as this CPM (continuous passive range of motion) machine are used for patients with injured limbs, for example.

frequently inconsistent, i.e., some items require more regular service and preventative maintenance than others.

- Seeking manufacturer assistance in the event of an issue: Failure of equipment or machinery causes a problem, particularly when the equipment was produced in another country, was made by a company that has gone out of business, or is unresponsive or sluggish to react to urgent technical support inquiries.
- Finding replacement components to repair failing equipment: Poor access to replacement parts, particularly for old equipment, is a common example.
- Maintaining a segregated collection of accessories and components: This is especially evident in hospital biomed, where a separate collection of accessories and parts is required for each type of monitor, anesthesia machine, etc.
- Persuading healthcare personnel and stakeholders that it is necessary to update equipment: Biomedical engineers often seek to persuade budget stakeholders that a piece of equipment is no longer safe, effective, or useful.

What caused the medical field to invite engineers into their workspaces?

In the 1960s, there was increased concern that patients could be put in serious danger because of some of the medical devices. This led to the the rise of the hospital electric safety industry and eventually to a new engineering discipline: clinical engineering. These clinical engineers work with hospital administration, doctors, and nurses; perform clinical research; and engage with patients, vendors, and regulation agencies as they relate to the use of hospital technical resources.

Can biomedical engineers engage in research on humans?

Yes, biomedical research can involve research with humans. This is called human subject research. The research must only be conducted by individuals who have received proper training, and they must follow generally accepted scientific procedures; and the human involved in the research must provide medical consent to participate.

What are some key moments in biomaterials history?

Biomaterials can be made of natural or synthetic materials and are used to support healing and restore functionality after disease or injury. Some materials include: metals, ceramics, clad, living cells, and tissue. Millions of lives have been saved because of advancements in the use of biomaterials. Here are key advances in biomaterials history:

- Ancient Egyptians used animal sinew to make surgical stitches
- In the 1860s, Dr. Joseph Lister developed the aseptic surgical technique that was used to prevent the spread of infection
- The first metal devices to fix bone fractures were used as early as the late-eighteenth to nineteenth centuries
- The first total hip replacement implant took place in 1938
- Polymers introduced for cornea replacements and blood vessel replacements in the 1950s and 1960s
- In 1961, synthetic materials were used for total knee replacements to help those with severe arthritis
- In 2002, soft contact lenses were introduced
- Around 2010, 3D printers were used to print patient-specific materials for life-saving applications

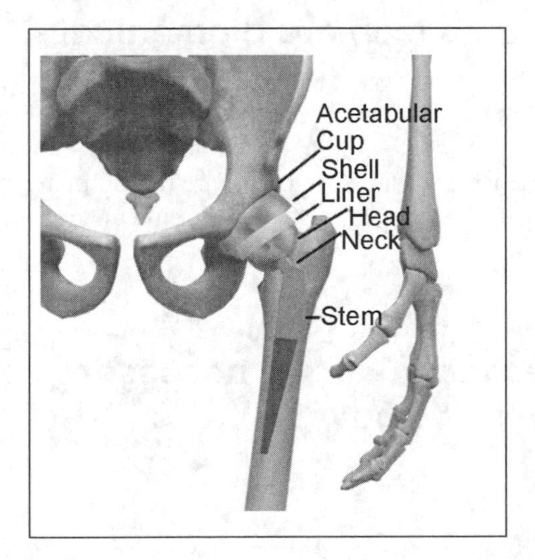

Hip replacement surgery has been around since 1938! Considerable advancements have been made since then, of course, thanks to engineers.

BIOMEDICAL ENGINEERING EDUCATION AND CAREER PREPARATION

How can someone become a biomedical engineer?

People interested in becoming a biomedical engineer should consider following these steps:

Step 1. High school: Students interested in biomedical engineering are required to take high school science courses, which include physics, chemistry, and biology, and math courses, which include geometry, algebra, calculus, and trigonometry. This is the minimum requirement for eligibility at any of the four-year colleges or universities nationwide.

Step 2. Undergraduate degree: Enroll in a four-year biomedical engineering program. For this admission, a student requires a high school diploma or GED along with good SAT or ACT scores.

Step 3. (optional) Graduate degree: While it is not a prerequisite to pursue a graduate degree, a student may choose to pursue a graduate-level education.

Step 4. (optional) Professional license: Although it is not mandatory, biomedical engineers with four years of pertinent work experience may apply for a professional engineering license.

How long does it take to become a biomedical engineer?

In general, it takes four years to complete a high school degree. Then, a bachelor's degree in biomedical engineering will take four years to complete. After this, most aspiring biomedical engineers take time to pursue either a graduate degree or professional certification.

What are some important qualities necessary for biomedical engineers to have?

- Analytical skills
- Critical thinking

DID YOU KNOW?

Are biomedical engineers supposed to go to medical school?

Biomedical engineering is somewhat correlated to the field of medical science, yet it is uncommon for a biomedical engineer to prepare for this career by going to medical school. According to reports from the U.S. Bureau of Labor Statistics, though, a few biomedical engineers who are interested in leading research in direct medical care applications do attend medical school.

- Communication skills
- Creativity
- Proficiency in science and technology
- Math skills
- Complex problem-solving skills
- Good decision making and judgment

Is graduate school-level education important in this field?

Graduate school education is an important part of biomedical engineering. Many engineering fields, like mechanical engineering or electrical engineering, do not require graduate-level training to get an entry-level job in their field, while biomedical engineering prefers or even requires graduate-level training.

For example, biomedical engineering-related professions usually require scientific research in fields such as pharmaceutical and medical device development; therefore, it becomes a must to have a graduate-level education.

Do biomedical engineering studies have an interdisciplinary nature?

Yes, they do. Graduate programs in biomedical engineering are highly varied. They sometimes emphasize collaboration with programs in other fields, such as medical school, or any other discipline in engineering.

Numerous subdisciplines exist in biomedical engineering: development of medical and therapeutic devices, orthopedic implants, biosignal processing and imaging, clinical engineering, etc.

Do biomedical engineers work with doctors?

Biomedical engineers typically design biomedical equipment and devices—for example, artificial internal organs—and machines for diagnosing medical issues and also provide technical support for biomedical equipment.

Thus, biomedical engineers handle medical problems through their research. Both doctors and biomedical engineers help people; however, they each do this in different ways. Biomedical engineers work on the design and improvement of equipment used to analyze and treat patients' medical conditions but not necessarily with patients themselves. Medical doctors directly impact patients, one at a time, through their work. Also, medical careers require advanced schooling, depending on the specialty.

What kinds of work environments are biomedical engineers in?

Work environments for biomedical engineers include clinics, hospitals, offices, and the research department or manufacturing department of companies and educational facilities.

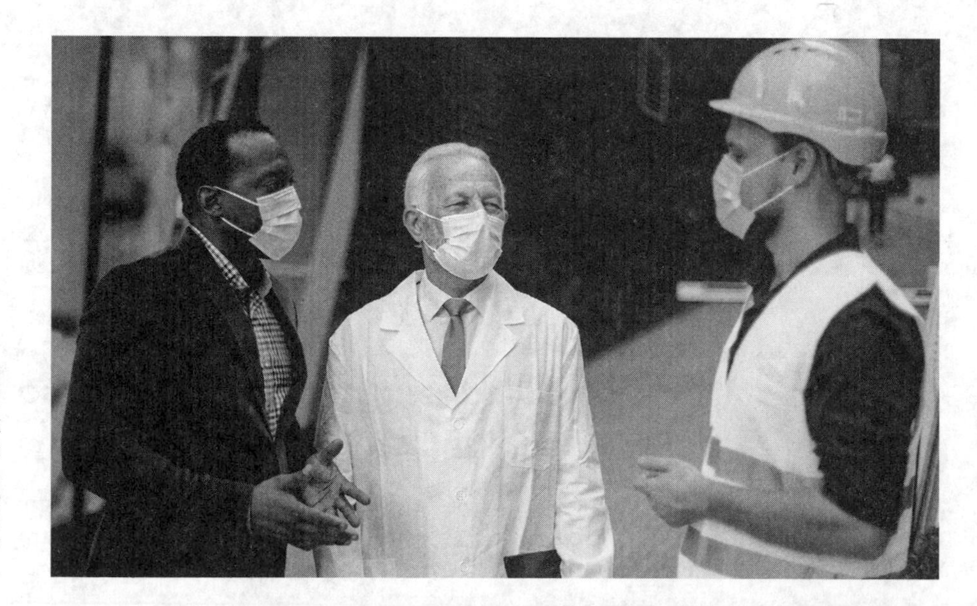

Doctors and biomedical engineers can work together on a project, naturally, but they do approach medical problems from very different angles.

What kinds of research opportunities are available in the field of biomedical engineering?

Biomedical engineering has a vast amount of research opportunities, as research is an integral part of the field. For example, research could be done in the field of biomaterials, which are important for implants, allergies, genes, medical imaging, and pharmaceuticals.

How does biomedical engineering contribute to society and people?

The output from biomedical engineering has a tremendous direct impact on the humanistic approach as well as advanced technology. The fundamental aim of biomedical engineering is to assist individuals through biomedical imaging, organ implants, bone implants, and so on. It works for individuals' well-being and their health.

What are example career options for biomedical engineers?

Biomedical Engineer—starting salary $44,000
Rehabilitation Engineer—starting salary $37,000
Clinical Engineer—starting salary $42,000
Bioengineering Research—starting salary $32,000

How much might a BME make?

The median annual wage for biomedical engineers was $92,620. The lowest 10 percent earned less than $56,590, and the highest 10 percent earned more than $149,440.

Top Industries That Hire Biomedical Engineers	Median Annual Wages for Biomedical Engineers
Navigational-, measuring-, electromedical-, and control-instrument manufacturing	$104,050

DID YOU KNOW?

Where in the United States are biomedical engineers paid the highest amount?

The U.S. Bureau of Labor Statistics reveals that Minnesota, New Jersey, Massachusetts, Arizona, and Connecticut are the top-paying states for biomedical engineering jobs. The highest-paid jobs in biomedical engineering are offered in California. Other well-paying areas include Austin, Texas, and Columbus, Ohio.

Top Industries That Hire Biomedical Engineers	Median Annual Wages for Biomedical Engineers
Research and development in the physical, engineering, and life sciences	$94,960
Medical equipment and supply manufacturing	$93,630
Health care and social assistance	$79,870
Colleges, universities, and professional schools: state, local, and private	$71,820

Is biomedical engineering a good career?

Numerous factors are considered when calling the engineering field a good major. Since biomedical engineering is an upcoming field and is multidisciplinary, it is based on the mainstays of all the traditional majors. For example, the creation of medical devices depends on the knowledge of electrical and mechanical engineering. Since it is forthcoming, work opportunities will increase in the coming years.

According to the U.S. Bureau of Labor Statistics, the work of biomedical engineers is anticipated to grow 6 percent by 2030, which is similar to the growth of all occupations. Biomedical designers likely will see employment growth as a result of expanding possibilities brought on by new advancements/technologies and applications to medical equipment. Biomedical engineering and traditional engineering programs—for example, mechanical and electrical—are typically good preparations for entering biomedical engineering occupations.

Can biomedical engineering be taken as a disciplinary course by international students?

Yes, it can. Like any other field, biomedical engineering applications are similar and require a GRE as well as a TOEFL score with relevant knowledge in this field during undergrad. The procedure for applying remains the same for this field, and furthermore, scholarships (grants) are available to students in this field.

Do scholarship opportunities exist in biomedical engineering?

Biomedical engineering is no different from other engineering fields. Many universities offer departmental scholarships specific to students in engineering programs; funds are available to currently enrolled sophomores, juniors, and seniors only.

Which are the best universities for a master's in biomedical engineering?

As per the student satisfaction, return on investment, and institutional excellence data from *U.S. News & World Report*, PayScale, IPEDS, and college websites, the following is a list of universities that provide the best master's degree programs in biomedical engineering:

1. Cornell University
2. Georgia Institute of Technology
3. Brown University
4. George Washington University
5. Duke University
6. Rice University
7. Johns Hopkins University
8. Colorado State University at Fort Collins
9. Carnegie Mellon University
10. Lawrence Technological University
11. Michigan State University

Are additional skills or more education needed by biomedical engineers?

An in-depth knowledge of the operational principles of the equipment (electronic, mechanical, biological, etc.), as well as knowledge about the application for which it is to be used, are necessary. For instance, in order to design an artificial heart, an engineer must have extensive knowledge of electrical engineering, mechanical engineering, and fluid dynamics as well as an in-depth understanding of cardiology and physiology. Designing a lab-on-a-chip requires knowledge of electronics, nanotechnology, materials science, and biochemistry. In order to design prosthetic replacement limbs, expertise in mechanical engineering and material properties as well as biomechanics and physiology is essential.

The critical skills needed by a biomedical engineer include a well-rounded understanding of several areas of engineering as well as the specific area of application. This could include studying physiology, organic chemistry, biomechanics, or computer science. Continuing education and training are also necessary to keep up with technological advances and potential new applications.

What professional societies can biomedical engineers belong to?

Because the United States as the largest biomedical engineering community in the world, there are many organizations that appeal to biomedical engineering and many of the sub- and cross-disciplines. These organizations include: the American Medical Informatics Association, IEEE Engineering in Medicine and Biology Society, Biomedical Engineering Society, Association for the Advancement of Rehabilitation and Assistive Technologies, and many others. In 1992, the American Institute for Medical and Biological Engineering was created to serve as the umbrella organization. Its purpose is to create unity across all the divisions and subdisciplines.

What research journals can provide more information about bioengineering?

- *Acta Biomaterialia* is an international journal that emphasizes the the relationship between biomaterial structures.
- *Advanced Healthcare Materials* is an international, interdisciplinary forum for peer-reviewed papers on high-impact materials, devices, and technologies for improving human health.

Engineering Square at Cornell University. The university's master's degree program in biomedical engineering was ranked number one in the United States.

- *Annals of Biomedical Engineering (ABME)* is an interdisciplinary, international journal that focuses on scientific research that include experimental and quantitative approaches.
- *Annual Review of Biomedical Engineering* has been in publication since 1999 and provides updates on significant developments in biomedical engineering.
- *Bioactive Materials* is an international, peer-reviewed journal that shares new information about bioactive materials.
- *Biofabrication* focuses on cutting-edge research regarding the use of cells, proteins, biological materials, and biomaterials as building blocks to manufacture biological systems and/or therapeutic products. It is also the official journal of the International Society for Biofabrication (ISBF).
- *Biomaterials* is an international journal that focuses on the science and clinical application of biomaterials.
- *Biomedical Engineering Education* is an interdisciplinary, international journal that presents articles on the practice and scholarship of education in bioengineering, biomedical engineering, and its allied fields.
- *Cardiovascular Engineering and Technology (CVET)* presents a wide spectrum of research, from basic to translational, in all aspects of cardiovascular physiology and medical treatment. It covers topics related to academic researchers and industrial researchers.
- *Cellular and Molecular Bioengineering (CMBE)* focuses on research that studies how cellular behavior arises from molecular-level interactions to tackle the challenge of improving human health.
- *IEEE Transactions on Medical Imaging (T-MI)* publishes research related to ultrasound, X-rays, magnetic resonance, radionuclides, microwaves, and optical methods. It emphasizes the common ground where instrumentation, hardware, software, mathematics, physics, biology, and medicine interact.

- *Medical Image Analysis* present new research in the medical and biological image analysis fields.
- *Nature Biomedical Engineering* is an online-only monthly journal that publishes highly significant research findings.
- *npj Regenerative Medicine* is an open-access, online-only journal dedicated to publishing the highest-quality research on ways to help the human body repair, replace, restore, and regenerate damaged tissues and organs.

KEYWORDS AND DEFINITIONS

What are some biomedical keywords and their associations to other terms and concepts?

- Biofabrication is associated with the generation of functional tissues and organs.
- Biointerfaces are associated with antimicrobial surfaces and coatings, mechanobiology and biocompatibility studies, interfaces for cell engineering and stem cell differentiation, 3D cell culture, and immunoengineering.
- Biomaterials are associated with nanomaterials, hydrogels, 2D materials, biopolymers, composites, biohybrids, biomimetics, and inorganic materials for biomedical applications.
- Devices for healthcare applications include diagnostics, wearables, implantable devices, microfluidics, BioMEMS, organs on a chip and labs on a chip, bioelectronics, biosensors, actuators, and soft robotics.

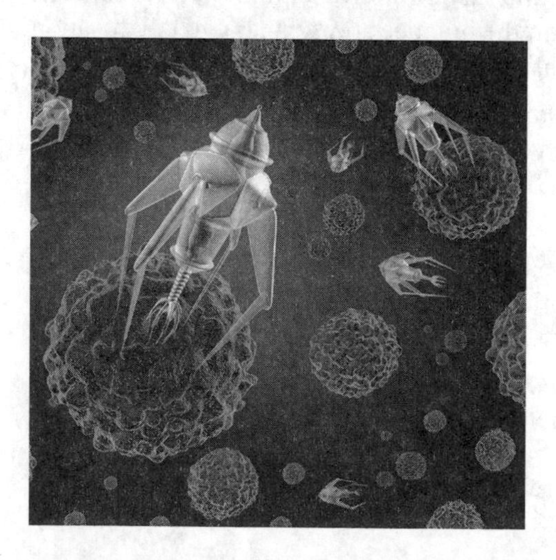

This illustration imagines nanobots that are designed specifically to kill cancer cells. In 2018, researchers did program nanobots to cut off blood from cancerous cells, but much work still needs to be done before nanobots become a viable cancer treatment.

- Nanomedicine and nanotechnology are associated with applications in drug delivery, imaging, theranostics, gene therapy, and immunotherapy; the therapy of infectious diseases, cancer, metabolic diseases, and cardiovascular diseases; and vaccines and precision medicine.
- Tissue engineering is associated with bone, ligament, and muscle tissue engineering; skin regeneration and wound healing; nerve grafts; cardiac patches; and tissue vascularization.

What is biomechanics?

Biomechanics is the study of the structure, function, and movement of the mechanical parts of biological frameworks at any level from whole organisms to organs, cells, and cell organelles utilizing the strategies for mechanics.

An article published by Study.com indicated that in the healthcare setting, biomedical research is concerned fundamentally with questions of how and why the musculoskeletal system behaves in the manner it does. Biomedical engineers utilize a variety of methods and principles, including traditional mechanics, physics, and mathematics. Given that this field of study is concerned principally with how and why the human body moves in specific ways, biomechanical experts will in general work in either research or product development.

The basic principles of biomechanics are as follows:

- Force
- Linked segments
- Impulse-causing momentum
- Stretch–shorten cycle
- Summing joint forces
- Continuity of joint forces
- Impulse direction
- Rotational motion
- Manipulating the moment of inertia
- Stress-causing strain

What is tissue engineering?

Tissue engineering is the combination of cells, engineering, and materials techniques with appropriate biochemical and physicochemical elements for the advancement of synthetic or natural human tissue in a laboratory, which can be used to help patients with a variety of medical conditions from severe burns

DID YOU KNOW?

What is biomechatronics?

Biomechatronics is an interdisciplinary area of the investigation of biology, mechanics, electronics, and control that centers around the research and development of assistive, therapeutic, and diagnostic devices and platforms that can respond to or even be utilized inside the human body for the treatment/compensation of patients who have some type of disability or disease.

Examples of biomechatronics research and technology include devices that permit neural control of prosthetics, implants that permit communication between prosthetic devices and the central nervous system, and exoskeleton devices that can be utilized to improve running.

Thus, research in this area helps to enhance the physiological functions of people with and without disabilities (for example, improving running speeds).

to organ failures; for example, tissues that have been effectively engineered include cartilage, skin, liver, and muscle tissue.

What is biomedical electronics?

Biomedical electronics is the part of bioengineering that is devoted to the advancement, design, and maintenance of devices that are utilized in healthcare settings such as medical clinics and centers. Examples of biomedical electronics include CT imaging systems, dialysis machines, emergency unit monitoring systems, and surgical lasers.

Experts of bioengineering will either work in the research and development of new platforms or in helping repair biomedical electronic equipment and managing appropriate use. Thus, the bachelor of science (BS) degree in biomedical electronics is designed for technicians who design and maintain hospital and other health-oriented electronics equipment.

What is bioinstrumentation?

Bioinstrumentation is an application of biomedical engineering that focuses on the devices and mechanics used to quantify, assess, and treat biological systems. It centers around the utilization of numerous sensors to monitor physiological qualities of a human or animal. It refers to high-tech, often expensive instrumentation used to conduct cutting-edge research in the biological sciences. Examples include instrumentation for imaging, disease diagnosis, and therapeutics.

What are biomedical sensors?

Biomedical sensors are an exceptional sort of sensors that provide the important interface between the biological environment and electronics systems under the strict requirements of having a high compatibility with the facilitating condition, being lightweight, and having a high reliability. Thus, biomedical sensors are devices that recognize explicit biological, chemical, or physical processes and, after that, transmit or report this information. These sensors are also often used to monitor the safety of medicines, food, environmental conditions, and other substances we may experience.

What are biochemical reactions?

A biochemical reaction takes place inside the cells of living things. It is the change of one molecule to a different molecule inside a cell. Biochemical reactions are mediated by enzymes, which are natural catalysts that can modify the rate and specificity of chemical reactions inside cells. The sum of all the biochemical

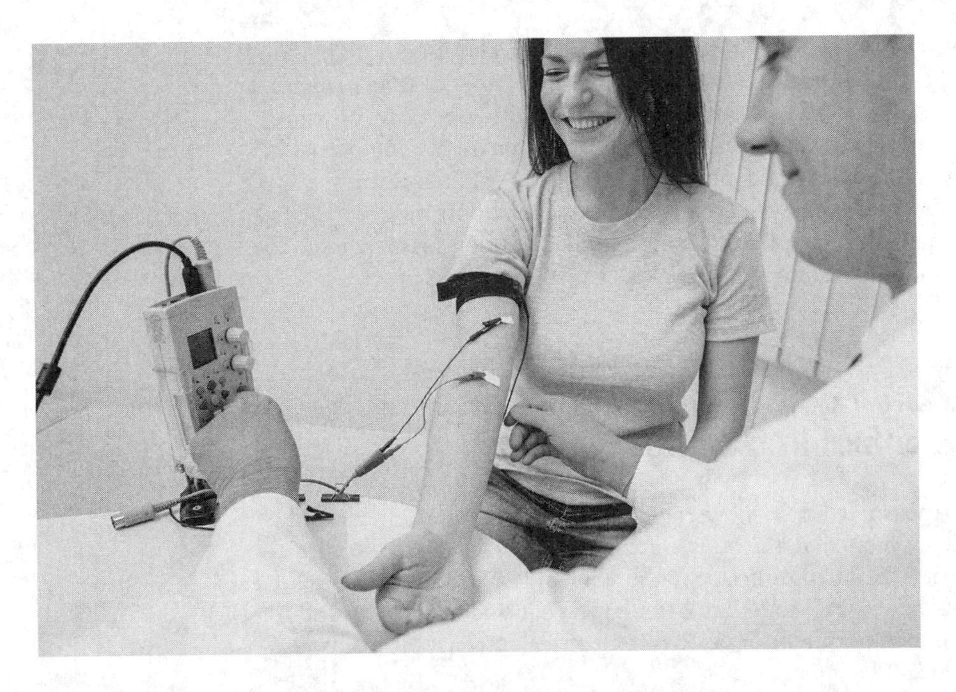

A patient is hooked up to a electromyography (EMG) sensor, which measures electrical signals traveling through the nervous system. An EMG is a type of biomedical sensor.

DID YOU KNOW?

What is bioinformatics?

Bioinformatics includes some core technologies and disciplines related to mathematics, statistics, and computational analysis to explore genetics instructions—DNA. This work can include creating databases that store information, artificial intelligence, developing algorithms to predict gene sequences and much more. One of the most common tasks in bioinformatics is to search databases for similar sequences. These databases are shared with researchers all over the world. One publically available tool is BLAST (http://www.ncbi.nlm.nih.gov/BLAST). This is a Basic Local Alignment Search tool that looks for similarities between genetic sequences of interest. The BLAST database is like a Google search engine for genetic sequences.

reactions in an organism is called a metabolism, which includes both exothermic and endothermic reactions. The four main types of biochemical reactions include oxidation-reduction, hydrolysis, condensation, and neutralization.

What is biosignal processing?

Our bodies constantly convey information about our well-being. This information can be captured through physiological instruments that measure blood glucose, heart rate, oxygen saturation levels, blood glucose, blood pressure, and brain activity. Generally, such measurements are taken at specific points in time and noted on a patient's chart. Doctors observe these readings as they make their rounds, and treatment choices are made dependent upon these readings. Thus, biosignal processing includes the examination of these measurements to give valuable information upon which clinicians can make decisions.

How does a biomedical engineer measure the electric potential in a body?

Biopotential measurements—how engineers determine and interpret the meaning of electrical potential in the body—are taken with electrodes. Noninvasive electrodes are attached to the surface of the skin, and invasive electrodes are placed under the skin. The electrodes are devices that conduct electricity and in medical purposes can either monitor or deliver electrical impulses in the body.

What are biomedical lasers?

Biomedical lasers are used for medical diagnosis and therapy. Nowadays, laser-based diagnostic devices are increasing in areas such as biomedical imaging and basic biological research. For example, the new field of optical tomography uses ultrashort laser pulses to recognize anomalies inside the body rather than depending on possibly unsafe ionizing radiation as other techniques do.

What is medical imaging?

Medical imaging refers to several different technologies that are utilized to see and observe the human body to analyze, monitor, or treat medical conditions. Each sort of technology gives distinctive information about the area of the body being examined or treated related to possible disease, injury, or the effectiveness of medical treatment.

What are some famous innovations in biomedical engineering?

- The first bionic arm, which was created in in 1993.
- Artificial organs, including kidneys and hearts, cardiac pacemakers, artificial skin, dentures, prosthetics, and hearing aids.

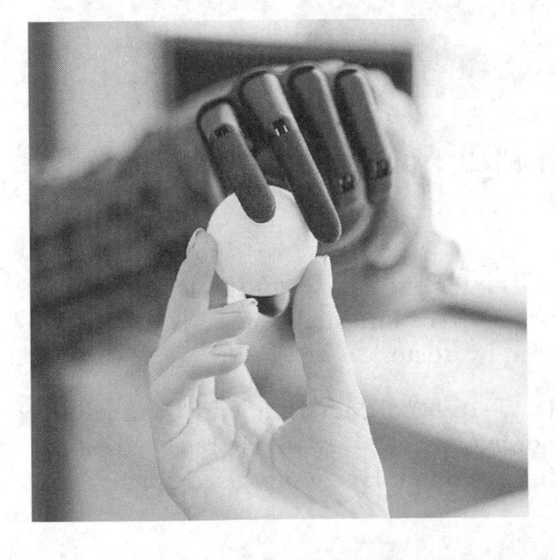

Bionic hands like this one are becoming more nimble and able to do such tasks as grip small items and even use a keyboard.

- Computer systems that help doctors analyze disease functions, automated insulin pumps, and corrective-surgery equipment and procedures.
- The development of a process of regeneration for spinal discs.
- The creation of vaccines for human use.
- The camera pill, which is a cost-effective scanning device that is the width of a fingernail and twice the length, can spot early signs of cancer and take quality color pictures.
- Robotic arms, which are also known as robosurgeons, are used to perform surgery autonomously on patients.
- Bionic legs are an exoskeleton that was developed to help people suffering from paraplegia; it is an easy-to-wear, artificially intelligent, bionic mechanism that can be used not only by patients in hospitals but also at home, allowing users to stand, walk, and even climb stairs.
- Bionic hands have three separate parts: the fingers, thumb, and palm, with each part featuring its own motorized control system; each finger is individually powered to allow users to grip multiple shapes.
- Bionic contact lenses allow wearers to see the world by superimposing computerized pictures onto their natural view; can be used by drivers and pilots to provide routes and information on weather or the vehicle or to help monitor a person's biological conditions, such as cholesterol level or the presence of viruses and bacteria. The collected data can then be sent wirelessly to a computer.
- An emotional–social intelligence prosthetic is a mind-reading machine that scans the brain signals of the user and alerts them to the emotional state of the individual they are in a conversation with. It may also be able to help people with autism improve the degree of learning the social cues of other people.

Who are some notable figures in biomedical engineering?

- Rene Laënnec (1781–1826) His interest in music and flute making led him to invent the stethoscope in 1816. The stethoscope is one of the most common instruments used by doctors.
- Benjamin Franklin Palmer (1847–1898) of Meredith, New Hampshire, invented the first artificial leg on November 4, 1846. He received patent number 4,834 for creating this invention. His design used springs and metal tendons.

- John Heysham Gibbon (1903–1973), an American surgeon, invented the first heart–lung machine.
- John Charnley (1911–1982) was instrumental in the development of the artificial hip replacement.
- Willem Johan Kolff (1911–2009) was known as the pioneer of artificial organs and the developer of the artificial kidney.
- Charles Hufnagel (1916–1989) developed an artificial heart valve and, along with his surgical team, developed a heart and lung pump for open-heart surgery.
- Wilson Greatbatch (1919 –2011) was an electrical engineer who helped to develop the first implantable pacemaker, which has helped a lot of people since the 1960s by giving them a second chance for life.
- Yuan-Cheng Fung (Bert, 1919–2019) is widely known as the Father of Biomechanics. He established the fundamentals of biomechanical properties in many of the different organs and tissues in the human body. In the field of tissue engineering, he studied remodeling, growth, and resorption of tissues, concentrating on doing these on blood vessels under stress in health and disease.
- Forrest Bird (1921–2015) was instrumental in the development of the mechanical ventilator.
- Wilhelm Roentgen's (1921–2015) work on cathode rays led him to the discovery of a new and different kind of rays, now known as X-rays.
- Michel Mirowski (1924–1990) invented the implantable cardioverter defibrillator.
- Thelma Estrin (1924–2014) is an early pioneer of bioinformatics. Through her research and publications, she documented how she used computers to map the brain.
- Graeme Clarke (1935–) was instrumental in the development of the cochlear implant.
- Robert Jarvik (1946–) developed the first artificial heart Jarvik-7. The first patient lived for an additional 112 days after the Jarvik-7 was implanted into his chest.
- Gordana Vunjak-Novakovic (1948–) is a pioneer in the tissue engineering field. She engineers tissues that can cure disease and heal illnesses. For example, her research team grows bone grafts that match a patients jawbone for use during reconstruction surgery.
- Chae-OK Yun (1963–) is a founder and CEO of GeneMedicine. Her company is a pioneer in oncolytic virus approaches in gene therapy.

Through her company she provided cutting-edge approaches to cancer treatment that can kill cancer cells without harming the normal tissues.

- James Collins (1965–) invented a number of devices and techniques, which included vibrating insoles for enhancing balance, a prokaryotic riboregulator, dynamical control techniques for eliminating cardiac arrhythmias, systems biology techniques for identifying drug targets, and disease mediators.

What are some awards given for biomedical engineering accomplishments?

The following is a list of awards given by the Biomedical Engineering Society (BMES):

- BMES Mid-Career Award: This award recognizes a member in good standing who has demonstrated significant leadership and achievements in biomedical engineering. These achievements may be in scholarship, education, mentorship, leadership, or the practice of biomedical engineering.
- Career Development Awards: These awards are available to graduate students, postdoctoral fellows, early career faculty, and early career professionals from underrepresented populations in biomedical engineering or involved in research and training focused on health disparities and minority health in biomedical engineering.
- Christopher Jacobs Award for Excellence in Leadership: This award honors leadership excellence in the field of cellular and molecular bioengineering (CMBE) and is named in memory and honor of Dr. Christopher R. Jacobs.
- Diversity Lecture Award: This award honors an individual, project, organization, or institution for impactful contributions toward improving gender and racial diversity in biomedical engineering. The award seeks to recognize lifetime achievements and high-impact activities (i.e., research, academia, and service) that are innovative and improve equity among biomedical engineering academia and industry members.
- Rita Schaffer Young Investigator Award: This award stimulates research careers in biomedical engineering.
- Robert A. Pritzker Distinguished Lecture Award: This award is BMES's premier recognition. Each year, the society bestows this prestigious award on individuals who have demonstrated impactful leadership and accomplishments in biomedical engineering science and practice.

What wages and benefits can a biomedical engineer expect?

DID YOU KNOW?

Wages for biomedical engineers are normally high; they offer above-average earning potential even to those who just graduated from college. The median wage in 2019 for biomedical engineers in California was $98,272 annually, or $47.25 hourly. The median wage for biomedical engineers in the Inland Empire area, a metropolitan region east of Los Angeles, is $67,251 annually, or $32.33 hourly.

THE FUTURE OF BIOMEDICAL ENGINEERING

What are the pros and cons of becoming a biomedical engineer?

Pros: Fast job growth, high salary, variety of workplace options (such as hospitals, universities, and research facilities), variety of job options (including manufacturing, educational, and medical), variety of specializations (such as biomaterials, biomechanics, and bioinstrumentation)

Cons: Working hours may sometimes vary, the minimum qualification required is a graduate degree, potentially hazardous exposure to fumes or radiation

What is the future of the biomedical engineering field?

Biomedical engineering is a field of science that unites engineering skills with applications in the medical division for its development and improvement and is one of the hottest careers around these days. From an economic perspective, medical diagnostics have progressively advanced, making triple in market value each year. Development and advancement in medical imaging and medical diagnostics are changing how medicine is practiced. Nowadays, with the help of research facilities, biomedical engineers around the globe have completely changed the way disease and trauma are dealt with, thus improving the quality of human life.

Eventually, the future of biomedical engineering is attached to the issues we face in fields like chemistry, materials science, and biology. An expanding utilization of the combination of medicine with innovation in technology for the treatment of diseases has occurred. In the present world, more and more doctors are utilizing the power of computers and other devices such as MRI scans and sonography devices. Biomedical engineers apply the ideas of science to create devices that are utilized in the treatment of diseases. Biomedical engineers work with other healthcare professionals, including doctors, medical caretakers, and specialists.

Computer Engineering, Computer Science, and Software Engineering

What is computer information science/computer science?

Computer science is the comprehensive study of computers and how they use computations and programming to solve simple and complex problems. Computer science has its roots in mathematics and engineering and encompasses the study of algorithms, data structuring and modeling, computer design, and information processing. Therefore, a computer scientist will use math and algorithms to make hypotheses, test theories, and perform high-level experimentation. Computer scientists often focus on several fields of study all centered around computers. For example, some areas of study include computer architecture, programming, software development, visualization techniques, artificial intelligence, and the design and maintenance of large databases and information systems.

What industries do computer scientists work in?

Computer science covers many different fields of study, and this work can be applied to various industries such as the following:

- Artificial intelligence (AI) and machine learning teaches computers and machines to perform tasks
- Business information technology applications address problems that businesses face
- Cybersecurity protects data, networks, and systems from online attacks and theft
- Data science uses a combination of statistics, mathematics, and engineering to analyze, visualize, and solve complex problems
- Digital content, such as product images and descriptions, social media fields, and promotional videos must be managed, organized, and stored
- Health informatics allow patients and doctors to manage a patient's healthcare information across several systems and platforms
- Human–computer interactions require the interactive environments made possible by computer science technology
- Library and information science
- Software engineering

What is computer engineering?

Computer engineering focuses on the practical use of computers. Computer engineering combines electrical and electronic engineering with computer science. A computer engineer can design and develop computer systems and

Computer engineers develop both hardware and software, combining knowledge of computer science and electrical engineering.

What is computer architecture?

Computer architecture is the process of selecting the hardware components to build a computer or system. A computer engineer must consider the technology available, the applications and domains, and the goals. Some applications and domains include desktop computing, servers, cell phones, embedded systems in automobiles and planes, and video game consoles.

DID YOU KNOW?

devices. For example, they may design, test, and improve computer hardware, such as computer circuit boards, and computer software, such as operating systems (i.e., macOS and Windows).

What does it mean for a computer to execute a task?

Execution is when the hardware follows the instructions of a specific program or code.

What is a database?

A database is a digital storage system that allows a user to organize, search, and retrieve data. An example of a database is an online clothing business that keeps track of product inventory and customer information. This might include how frequently a customer visits the website, where the customer is from, and what items they look at and/or purchase.

How are databases used by computer scientists?

Computer scientists created databases for streaming services like Netflix and Disney+ to keep track of what the viewer watches and what shows are available to stream so that they can make recommendations on what to watch next. With live gaming across the world on different platforms, a large amount of user data is created. Players' information must be gathered and given to other players on demand. Computer scientists also create and manage databases in finance, health care, cloud storage, sports, government, social media, and weather.

What is an operating system?

The operating system (OS) is the most important software on a computer because it manages the interactions between all the other software and the hardware. The user interfaces with system and application software, and the operating system interfaces directly with the hardware (CPU, memory, input/output) and the system.

What is an example of an operating system?

Some common operating systems include Microsoft Windows, macOS, and Linux. The Linux system is often a favorite among developers and computer engineers who like to customize their computers to meet specific needs. Microsoft Windows and macOS allow very minimal changes and come preloaded on their respective PC and Mac computers.

What is logic?

Logic is the science of reasoning; it is one of the oldest disciplines in history and has been studied throughout time. It is used across many areas of our lives; humans use logic to state observations, create definitions/make conclusions, and create theories that explain how things work. Mathematicians use logical proofs to convince themselves and others that the conclusions we made about how things work are true. Alan Turing is one such mathematician who began to think about how logic can be used by computers to make decisions and perform tasks. Computer logic is a tool that computer programs use to describe the world and is also used for creating artificial intelligence and simulations.

What is the difference between a web developer and a website designer?

Web development and web design are two very exciting careers that are connected yet different. A web developer is someone who creates, builds, and performs routine maintenance of the core structure of a website, while a web designer is responsible for the design of the layout, appearance, and overall usability of the website.

Do computer scientists make robots?

Computer scientists do not make robots or automated devices. They write the software and programs that are used to control robots.

HISTORY

What is the old definition of a computer?

Did you know that until about 1950, the word "computer" referred to someone who performed mathematical calculations? The inventors of the early version of the computer we use today had to develop a device that did not just perform mathematical calculations. It had to be able to solve problems and make decisions. This is what led to the invention of programming.

What is the earliest calculator?

The abacus dates back to at least 1100 B.C.E. It was widely used by merchants and traders on the continents of Asia and Africa. Today, it can still be

The abacus has been used for milennia to make quick calculations, and some people still use it today in many Asian countries.

used for counting, addition, subtraction, multiplication, division, fractions, counting money, and even more complex mathematics. For example, when using it for counting, each row represents a unit/placeholder, and each bead has a value. The bottom first row represents the ones place, the second row represents the tens place, the third row represents the hundreds place, and the fourth row represents the thousands place; each additional row above is multiplied by 10. In the first row are 10 beads. Each bead has a value of 1; in the second row, each bead has a value of 10, and in the third row, each bead has a value of 100. So, counting to 11 would require one bead from the second row and one bead from the first row. The beads are moved from one position to another and represent specific values.

Who are some noteworthy computer scientists?

Throughout history, several computer scientists have made significant contributions to the field of computers. Some of these scientists are:

- John von Neumann (1903–1957): A Hungarian American known for developing the field of game theory.

A detail of the statue of Alan Turing located at the University of Surrey in England.

- Alan Turing (1912–1954): An English computer scientist who has been called the Father of Computer Science because of his early and pivotal role in helping the British government to defeat the Germans in World War II by developing a computer that could decipher the code being used by the Germans, thus giving the British forces and their allies an advantage.
- Joseph Carl Robnett Licklider (1915–1990): He is known for his invention of cloud computing. Cloud computing has become very popular in more recent times with companies such as Google, Amazon, Microsoft, and Apple, which have adopted this technology to offer storage capabilities to their users at fairly low costs and with high security measures.
- John McCarthy (1927–2011): Known for his invention of artificial intelligence (AI), McCarthy also invented Lisp, which is the programming language most commonly used in AI applications.
- Dennis Ritchie (1941–): One of the creators of the C language, the most common programming language used in computer systems. Dennis is also known for his cocreation of Unix, the programming language used by Apple in its operating system, MacOS.
- Brian Kernighan (1942–): The cofounder of the C language and Unix.
- Steve Wozniak (1950–): Known for his development of Apple. Wozniak is referred to as the technical genius behind Apple through his superior and unique programming style and skill.
- Tim Berners-Lee (1955–): Known for developing the World Wide Web.
- James Gosling (1955–): Known for his invention of the Java programming language. Like C programming, Java is very popular in software development and is used in the design of many software systems.

One of the giants of the computer era is Steve Wozniak, the cofounder of Apple, Inc.

- Mark Dean (1957–): Co-creator of the IBM personal computer that was released in 1981. He also contributed to the development of the color PC monitor and the first gigahertz chip. Mark became the first African American person to be named an IBM fellow.
- Kimberly Bryany (1967–): Founder of Black Girls code in 2011, which is a nonprofit organization that teaches programming to young girls of color.
- Linus Torvalds (1969–): The founder of Linux, which is the core of many PC operating systems. He is also known for his creation of Git, which is another system used frequently in software development.

SOFTWARE ENGINEERING

What is software?

Software is a set of logical instructions that carries the predefined tasks that a computer needs to do in order to perform an assignment. The software will tell a computer what it needs to do and how it needs to do it. The term "software" also refers to the computer programs, such as Microsoft Word, that are loaded onto a computer; software, therefore, can refer to either system software or application software. System software is responsible for the core functionality of the computer. Application software is customized to the needs of the user.

What is software engineering?

Software engineering includes the process of using applicable knowledge about computing systems and requirements to create valid and reliable software systems that can be used in various contexts. Software engineering also involves applying engineering principles to design, build, and test software products that meet users' needs. A software engineer understands how to write software and also how the software will interact with the computer's hardware.

What are the characteristics of good software products?

- Maintainability: Software should be robust enough to allow future programmers to make changes

- Correctness: Software must meet the user's requirements
- Reusability: Software should be created in a way that allows other programmers to use it
- Reliability: Software should fail as little as possible
- Portability: Software should be able to be used on other computer configurations
- Efficiency: Software execution should be as efficient as possible and take up as little space as possible

What does a software engineer do?

A software engineer meets with clients to understand what they need the software to do. Once they know what the user's software needs are, the engineer will decide which computer language they will use and how the programming language will interact with the computer hardware. Some software engineers specialize in the performance of a software application that has already been created. Other software engineers write new software in order to check the quality of another software product. Another group of software engineers acts as hackers to check the security of a software program.

What are the subfields of software engineering?

Software engineering has several subfields. It is a quickly evolving field, and these engineers are needed in many capacities, including these:

- Management: Oversees a software project team
- Architecture: Designs and selects the framework, which includes the programming language and required program functions
- Embedding: Writes code for software that is used by robots
- Modeling and simulation: Uses software to build a virtual environment to figure out how a system, process, or product will behave
- Entertainment systems: Designs, tests, and builds the software programs used on smart devices like TVs and game systems
- Networking: Creates software programs that allow computers to communicate with each other and fixes problems with computers that cannot communicate with each other

Who is Grace Hopper?

Grace Brewster Hopper (1906–1992) was a U.S. computer pioneer, one of three modern programmers, and a naval officer. She earned a master's degree in mathematics in 1930 and later a Ph.D. in the same area of study. Hopper became a professor and later took a leave of absence to join the U.S. Naval Reserve. She was assigned to the Bureau of Ships Computation Project, which was based out of Harvard University. She worked with Howard Aiken (1900–1973), who developed the Mark I computer, one of the earliest recognized electromechanical computers. Hopper wrote the programs for Mark 1, plus a 561-page user manual. She remained in service while based out of Harvard University and helped to develop the Mark II and Mark III computers. She became the first person to refer to computer problems and coding errors as "bugs." She died with the distinction of being the oldest serving officer in the U.S. Armed Forces. In 2016, Grace Hopper was awarded the Presidential Medal of Freedom, the highest civilian honor, in recognition of her contributions to computer science.

What are the current trends and terminologies in software engineering?

- Automation in manufacturing plants

Rear Admiral Grace Hopper is shown here in her Washington, D.C., office in 1978. She was a pioneer in computer programming, devising the programming language that later became COBOL.

- Developing a program that can make sense of a large amount of customer data
- Artificial intelligence (AI) and machine learning (ML)
- Cloud-based data storage systems
- Internet of Things (IoT)-connected devices—think cars and home appliances—and technologies will need a software platform that supports them. The hardware of a computer is tangible, and the parts of a virtual machine are software code. A computer scientist makes a virtual version of the computer that behaves like an actual computer but doesn't interfere with the physical computer's main functions.

What are some examples of software development processes?

Software development can be broken into distinct stages with specific activities.

- The waterfall approach follows specific steps with an emphasis on planning, budgeting, and implementation. This is a classic approach and is not flexible. Some people remember the steps using the mnemonic A Dance in the Dark Every Monday, representing the words Analysis, Design, Implementation, Testing, Documentation, Execution, and Maintenance.
- The prototyping approach can be used throughout the development process to provide the client with incomplete versions of the software for testing throughout the development process.
- The spiral approach combines some waterfall and rapid prototyping elements and focuses on understanding and minimizing risks. The developer takes multiple trips around the spiral, performing the same tasks each time on a new or revised prototype.

Can software testing be done in different ways?

Yes, software testing can be done in several ways. Software testing is important because it is the means by which developers can assess if the developed software can execute the tasks it was designed to do and if it works as it is expected to. By testing software at various points in the design process, developers are able to evaluate how the software will function under certain conditions as well as across the multiple types of computer platforms and operating systems that currently exist. Software testing tends to fall under two broad categories: functional testing and nonfunctional testing.

Functional testing is performed when the goal is to assess how well the software meets the needs of the user or the customer. This involves measuring the software against all of the end user's requirements. Some common tests that are conducted as part of functional testing are unit testing, integration testing, system testing, and acceptance testing.

- Unit testing: This is often the first test done on any software, and its goal is to check the accuracy of the code used to write every unit of the software. At this stage of testing, the intent is to check how well the software functions overall under the most basic use.
- Integration testing: This test is performed after all units of the software have been tested. Integration testing is done on the software as a whole to see how the units function as an integrated system. The goal of this test is to ensure that the interactions between the units are flawless and harmonious.
- System testing: This is also called a black-box test. The goal of this test is to ensure that the software overall meets specific requirements. At this stage of testing, the software is treated as a fully functional and integrated system, and the system test is done to evaluate how the software will perform under all possible user conditions.
- Acceptance testing: This test is performed last because its goal is to ensure that the final software design is in accordance with all the user requirements. At this testing stage, the software is ready to be deployed and used.

Nonfunctional testing is performed to assess the operational aspects of the software, meaning that the software is verified to ensure that it is fit for use and that it will stand up to various kinds of operations. Some of the most common types of nonfunctional tests are performance testing, security testing, usability testing, and compatibility testing.

- Performance testing: This type of test is used to ascertain the behavior of the software under different operating conditions. Here, the software's ability to respond to various user conditions is assessed through different authentic conditions, such as increasing load on the system, stress—i.e., how will the software perform at or beyond its peak load—endurance testing, and spike testing.
- Security testing: This test is performed to ensure that any data stored or used with the software is kept and managed in a safe way. This type of testing is necessary for all software given the prevalence of cyberattacks and the increased use of cloud storage services.
- Usability testing: This test is often conducted to assess the end user's experience with the software. At this stage, the overall look and feel of the software is evaluated to see how intuitive the system is as well

as look for any extraneous steps that may impact the overall flow of the system.

- Compatibility testing: This final type of nonfunctional testing is performed to ensure that the software will operate as intended across multiple computer platforms, operating systems, or various work environments. At this stage, the goal is to ensure that no matter who the user is, where they are located, or what type of computer system they have, they will have the same experience using the software.

COMPUTER HARDWARE, NETWORKS, AND SYSTEMS

What is computer hardware?

Computer hardware is the group of physical components that a PC or laptop requires to perform its functions. The most commonly discussed computer hardware parts are the hard drive, central processing unit, keyboard, monitor, speaker/microphone assembly, motherboard, mouse, graphics card, sound card, memory card, battery, and case/housing. In newer-model computers, some of these parts are integrated into the motherboard. However, whether these are separate parts or integrated into one piece of hardware, i.e., the motherboard, the speed of a PC or laptop is determined by the quality of the computer hardware.

What is a motherboard?

The motherboard is one of the most important pieces of hardware; it acts like a hub for the computer. It is the main piece of circuitry that connects to all the other components. It essentially allows all the other components to communicate with each other. Some of the hardware elements that it connects include the power supply, hard drive, central processing unit, ventilation fans, web camera, and graphics card. When building a personal computer, many people will purchase motherboards that have room to expand in order to include more components.

What is a central processing unit (CPU)?

The central processing unit is a circuit board that receives, reads, and executes instructions at a very rapid pace. Some people think of the CPU as the computer's brain. Old computers used to have to be physically wired and re-

Motherboards like this one contain the circuitry that connects all the components of the computer.

wired in order to perform specific tasks. Modern computers that use a CPU now use circuitry in order to execute instructions.

Why are microprocessors important?

A microprocessor is a very small circuit board that fully contains the central processing unit. It receives, reads, and executes instructions from a program and performs calculations. It has billions of transistors that allow it to perform billions of calculations at a rapid speed. Microprocessors can be found in video games, cell phones, computers, and cars. They are important because their size allows them to be used in many different types of applications, and they are very fast and efficient at performing complex tasks. Here are some examples of microprocessors:

- Memory chips: Random-access memory (RAM), read-only memory (ROM), and nonvolatile rewritable flash memory
- Data-storage devices: Hard discs, solid-state drives, and optical drives
- Input devices: Keyboards, mice, joysticks and gaming controllers, cameras, microphones, scanners, touch screens, and remote sensors
- Output devices: Printers, monitors, audio devices, and remote controls
- Networking components: Adapters, modems, switches, and routers

What types of large systems do computer engineers design?

Many modern systems in the world require computers to perform their jobs. This includes the internet, wireless networks, the computer systems that

What are some everyday devices that computer engineers create?

DID YOU KNOW?

When we think of computer engineering, it is normal to imagine that these engineers only work on high-level, elaborate technology products. While this is true, some computer engineers also work on the simple, everyday devices we interact with. Some everyday devices that computer engineers had a role in developing include video game platforms, laptop and desktop computers, and cell phones. Computer engineers helped to design and develop several appliances that can be found around the home, such as refrigerators, washing machines, televisions, dishwashing machines, and even some electronic toys. Most newer vehicles have completely computerized systems that alert us when something is wrong by means of sensors and alert lights. These are all connected to the computer system of the car.

are integrated into spacecraft and automobiles, and even modern processing and manufacturing plants.

What is a computer network?

When two or more computers are connected for the purpose of sharing resources and data or for extending the digital capabilities of a system, they are called a network. Computers can be connected within the same location or across a wider geographical area. Computer networks can be either wired or wireless. The two main types of computer networks are local-area networks (LANs) and wide-area networks (WANs).

In a local-area network, LAN, computers are connected within a fairly small physical area. This can be within a single school building, a larger office building, or a compound, as in across a university. Computers in a LAN are often connected using physical wires, such as ethernet or fiber-optic cables, or through wireless internet connections. LANs do not only consist of computers but may also include printers, file servers for storing large amounts of data, and other peripheral devices relevant to the context in which the LAN exists. In a LAN, all users are able to access common files and print to the connected printer(s), no matter where they are located. LANs also enable the shared use of programs.

A wide-area network, WAN, as the name suggests, connects computers across much larger networks that may span different towns, cities, countries, and, in some cases, even continents. This type of network is often connected

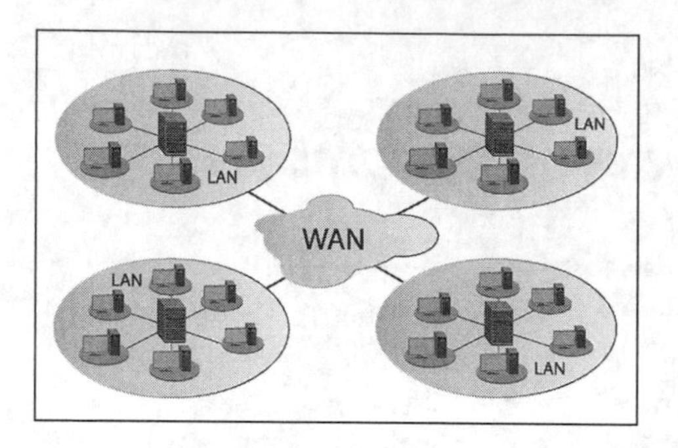

Figure 16: A Wide Area Network (WAN) will connect to multiple Local Area Networks (LANs) to efficiently cover large geographical areas.

using telecommunication links such as telephone lines and satellites. This type of network configuration is important for multinational companies or government entities that have facilities in various locations and sometimes countries.

Other types of networks that are not as widely discussed but are still relevant to some people include:

- Personal Area Network (PAN): As the name suggests, this type of network is limited to an individual workspace where the other devices (cell phone, printer, headset, smart watch, etc.) owned by the individual are connected, wired or wireless, to their computer.
- Home Area Network (HAN): This type of network exists within a home environment where devices—e.g., computer, television, cell phones, printers—are all connected.
- Campus Area Network (CAN): Very similar in structure to a LAN or maybe an integration of several LANs within a certain area. Campus networks are common at universities, large government agencies, or business corporations.
- Metropolitan Area Network (MAN): Like a campus network, MANs may consist of several LANs connected within a certain metropolitan area, as in a city.
- Enterprise Private Network (EPN): This type of network is more common among large corporations where the goal is to ensure that employees at different locations can tap into and share the same set of resources.
- Global Area Network (GAN): This characterizes a worldwide network that connects several other types of networks all over the world. The internet is a classic example of a global area network.

What are embedded systems?

Embedded systems use hardware to collect data and programming to make sense of the data that the hardware collected (consider a temperature sen-

sor that reads the value of the ambient temperature; the program could interpret that temperature as hot or cold). When considering a cell phone, a computer engineer focuses on circuit chip design, memory, microprocessor size and selection, and control. A computer scientist focuses on the operation system, code for all the applications, and safety aspects such as encryption.

What is the difference between computer science and information technology?

The disciplines of computer science and information technology are similar in that they are both technical and require significant amounts of knowledge about computers and how computing systems operate as well as technical skills to design and create various computer-related systems in order to meet certain needs. For instance, the computer scientist would be the person who designs the computer network, while the information technologist would be the person to constantly monitor the system to ensure that it functions like it should. Furthermore, several fundamental differences exist between these two disciplines. For example, computer science tends to focus on the design, creation, and testing of new computer programs, operating systems, and computer applications. When someone studies computer science, they specialize in the use of programming languages, the construction and deconstruction of algorithms, and the application of computer theory. To this end, someone with a degree in computer science typically works as a software engineer or software developer.

Information technology, on the other hand, is more concerned with the business side of technology. Someone trained in information technology will seek to design and create solutions to address any number of business problems. Someone with a degree in information technology would work as a systems analyst, IT consultant, network developer and administrator, computer support specialist, or data analyst.

Who is Frances Allen?

Frances Elizabeth "Fran" Allen (1932–2020) was a pioneer in computer engineering. She was born in Peru, New York, to a father who was a farmer and a mother who was a schoolteacher. She was inspired by her math teacher to become a math teacher. She earned her first degree in mathematics with a minor in physics from New York State College for Teachers (now the State University of New York at Albany). She then worked as a high school mathematics teacher. She hoped to become fully certified as a teacher and went to the University of Michigan, earning her master's degree in mathematics. During her studies, she

A computer scientist for IBM and winner of the Turing Award, Frances Allen pioneered the field of optimizing compilers.

took computing courses and learned how to program. Based on her experience, IBM offered her a position as a programmer, and she joined the company in 1957. She had the job of learning the FORTRAN programming language and then teaching IBM research scientists how to use it. As she prepared for this training, she read the source code for the FORTRAN compiler. This launched the direction of the rest of her career. She was a leader in compiler development and computing.

What are computer systems?

A computer system is a combination of unique devices (hardware) and programs (software) designed to perform a specific function. A typical computer system consists of hardware devices that perform input, output, processing, and storage functions.

- Input devices are used to enter data or instructions into the computer so that it can perform specific tasks. Some examples of input devices are keyboards, mice, trackballs, scanners, light pens, modems, card readers, and microphones.
- Output devices are used to display information that is the direct result of what the computer does. Some examples of output devices are monitors, printers, plotters, speakers, and earphones.

How do computer scientists use math to solve problems?

Mathematics is a very important aspect of computer science. Since computer science is concerned with the principles of using computers to process information at extremely high and fast levels as well as designing hardware and programming software, computer scientists must have a highly functioning knowledge of algebra, statistics, calculus, and binary and discrete mathematics.

DID YOU KNOW?

- Processing devices: The central processing unit (CPU) is the main and most important portion of a computer, which takes the data and instructions inputted into the computer, performs a set of manipulations, and returns an output.
- Storage devices are used to store data and information that are produced by the computer or needed by the computer to perform certain functions, such as program files. Some examples of storage devices include internal and external hard drives, USB drives, and cloud storage.

A computer system also consists of application and system software programs that are a set of instructions that tells the computer what to do.

What is an algorithm?

An algorithm is a sequence of instructions to perform a specific task. Each task can have many different algorithms. An algorithm is not a computer program.

Algorithm: to determine the largest of three numbers.

1. Start
2. Read three numbers A, B, C
3. Find the larger number between A and B and store it in MAX_AB
4. Find the larger number between MAX_AB and C and store it in MAX
5. Display MAX
6. Stop

VIRTUAL MACHINES

What are virtual machines?

A virtual machine is similar to a physical laptop, desktop computer, or other similar device. The hardware of a computer is tangible, and the parts of a virtual machine are software code. Basically, a computer scientist makes a virtual version of the computer that behaves like an actual computer but doesn't interfere with the physical computer's main functions.

Why is a virtual machine used?

Virtual machines have diverse uses. They can be used to build and launch software applications. Computer scientists develop operating systems and use virtual machines to test the operating systems in similar environments. Additionally, a computer scientist could use virtual machines to improve existing virtual environments. It is important to back up a computer; in some cases, the user will use the virtual machine to back up their work and operating system preferences so that in the case of failure, they can retrieve the backed-up version. Virtual machines are safe environments to run data that might be infected with a virus because it will not affect the physical computer. Finally, virtual machines can be used to run software that cannot be run on a physical computer's operating system.

Is it possible to have multiple virtual machine environments?

Yes; a user can run multiple virtual machine environments on one computer using a piece of software called a virtual machine manager.

What are the benefits of using virtual machines?

Running virtual machines has many benefits, including cost savings of running multiple environments on one computer, speed of booting up, very portable and easy to move, ability to add software apps across environments, and protection of the host computer from apps that may have viruses.

PROGRAMMING

What is a computer program?

A computer program is a set of instructions, written in a specific language, that must be executed to accomplish a task. It is the program that instructs the computer to take an input, manipulate it in some way, and return it to the user in an output that they can comprehend. Some common computer programs are operating systems such as Microsoft Windows and MacOS and web browsers such as Google Chrome, Mozilla Firefox, Safari, and Microsoft Edge. Other commonly used programs are word processors such as Microsoft Office Suite, video games, and data manipulation programs. Computer programs are stored on the hard drive of the computer as a series of zeroes and ones in a file. At the time of operation when the user opens the file associated with the program (the input), the central processing unit reads these zeroes and ones in the file and then tells the computer what to do. The output of this process is then returned to the user in the form of pictures or information based on what the program is designed to do.

What is a programmer?

A computer programmer is a person who writes programs that instruct the computer, application, software, or website. The programmer is also responsible for designing new programming and updating programs. Computer programming is also informally called "coding."

Who was Ada Lovelace?

Ada Lovelace (1815–1852), whose full name was Ada King, Countess of Lovelace, was a mathematician from England who was considered to be the first computer programmer. Her father was a poet, and her mother was a mathematician who insisted that her daughter also learn mathematics. Early in her career, she translated works by other mathematicians and engineers and was introduced to an early prototype of computers developed by Charles Babbage. Although Charles Babbage created the machine, Ada Lovelace could better communicate how the machine functioned. She wrote articles to describe how the calculating machine worked.

Considered the first computer programmer, Ada King, Countess of Lovelace, wrote code for Charles Babbage's analytical engine in 1842.

Diagram for the computation by the Engine of the Numbers of Bernoulli. See Note G. (page 722 *et seq.*)

Number of Operation.	Nature of Operation.	Variables acted upon.	Variables receiving results.	Indication of change in the value on any Variable.	Statement of Results.	Data			Working Variables.										Result Variables.			
						$^1V_1$ 0001 [1]	$^1V_2$ 0002 [2]	$^1V_3$ 0004 [n]	$^0V_4$ 0000	$^0V_5$ 0000	$^0V_6$ 0000	$^0V_7$ 0000	$^0V_8$ 0000	$^0V_9$ 0000	$^0V_{10}$ 0000	$^0V_{11}$ 0000	$^0V_{12}$ 0000	$^0V_{13}$ 0000	$^1V_{21}$ B_1 in a decimal fraction. [B_1]	$^1V_{22}$ B_3 in a decimal fraction. [B_3]	$^1V_{23}$ B_5 in a decimal fraction. [B_5]	$^0V_{24}$ 0000 [B_7]
1	×	$^1V_2 \times {}^1V_3$	$^1V_4, {}^1V_5, {}^1V_6$	$\{^1V_2 = {}^1V_2;\ ^1V_3 = {}^1V_3\}$	$= 2n$	...	2	n	$2n$	$2n$	$2n$											
2	−	$^1V_4 - {}^1V_1$	$^2V_4$	$\{^1V_4 = {}^2V_4;\ ^1V_1 = {}^1V_1\}$	$= 2n - 1$	1	...	...	$2n-1$													
3	+	$^1V_5 + {}^1V_1$	$^2V_5$	$\{^1V_5 = {}^2V_5;\ ^1V_1 = {}^1V_1\}$	$= 2n + 1$	1	...	...	...	$2n+1$												
4	÷	$^2V_5 \div {}^2V_4$	$^1V_{11}$	$\{^2V_5 = {}^0V_5;\ ^2V_4 = {}^0V_4\}$	$= \frac{2n-1}{2n+1}$	...	...	...	0	0	...	...	...	...	...	$\frac{2n-1}{2n+1}$						
5	÷	$^1V_{11} \div {}^1V_2$	$^2V_{11}$	$\{^1V_{11} = {}^2V_{11};\ ^1V_2 = {}^1V_2\}$	$= \frac{1}{2} \cdot \frac{2n-1}{2n+1}$	...	2	...	...	...	...	...	...	...	...	$\frac{1}{2} \cdot \frac{2n-1}{2n+1}$						
6	−	$^0V_{13} - {}^2V_{11}$	$^1V_{13}$	$\{^2V_{11} = {}^0V_{11};\ ^0V_{13} = {}^1V_{13}\}$	$= -\frac{1}{2} \cdot \frac{2n-1}{2n+1} = A_0$	...	...	...	...	...	...	...	...	...	...	0	....	$-\frac{1}{2} \cdot \frac{2n-1}{2n+1} = A_0$				
7	−	$^1V_3 - {}^1V_1$	$^1V_{10}$	$\{^1V_3 = {}^1V_3;\ ^1V_1 = {}^1V_1\}$	$= n - 1\ (= 3)$	1	...	n	...	...	...	...	...	...	$n-1$							
8	+	$^1V_2 + {}^0V_7$	$^1V_7$	$\{^1V_2 = {}^1V_2;\ ^0V_7 = {}^1V_7\}$	$= 2 + 0 = 2$	...	2	...	...	...	...	2										
9	÷	$^1V_6 \div {}^1V_7$	$^3V_{11}$	$\{^1V_6 = {}^1V_6;\ ^0V_{11} = {}^3V_{11}\}$	$= \frac{2n}{2} = A_1$	...	...	...	...	...	$2n$	2	...	...	...	$\frac{2n}{2} = A_1$						
10	×	$^1V_{21} \times {}^3V_{11}$	$^1V_{12}$	$\{^1V_{21} = {}^1V_{21};\ ^3V_{11} = {}^3V_{11}\}$	$= B_1 \cdot \frac{2n}{2} = B_1 A_1$	...	...	...	...	...	...	...	...	...	...	$\frac{2n}{2} = A_1$	$B_1 \cdot \frac{2n}{2} = B_1 A_1$	....	B_1			
11	+	$^1V_{12} + {}^1V_{13}$	$^2V_{13}$	$\{^1V_{12} = {}^0V_{12};\ ^1V_{13} = {}^2V_{13}\}$	$= -\frac{1}{2} \cdot \frac{2n-1}{2n+1} + B_1 \cdot \frac{2n}{2}$	...	...	...	...	...	...	...	...	...	...	....	0	$\left\{-\frac{1}{2} \cdot \frac{2n-1}{2n+1} + B_1 \cdot \frac{2n}{2}\right\}$				
12	−	$^1V_{10} - {}^1V_1$	$^2V_{10}$	$\{^1V_{10} = {}^2V_{10};\ ^1V_1 = {}^1V_1\}$	$= n - 2\ (= 2)$	1	...	...	...	...	...	...	...	...	$n-2$							
13	−	$^1V_6 - {}^1V_1$	$^2V_6$	$\{^1V_6 = {}^2V_6;\ ^1V_1 = {}^1V_1\}$	$= 2n - 1$	1	...	...	...	...	$2n-1$											
14	+	$^1V_1 + {}^1V_7$	$^2V_7$	$\{^1V_1 = {}^1V_1;\ ^1V_7 = {}^2V_7\}$	$= 2 + 1 = 3$	1	...	...	...	...	...	3										
15	÷	$^2V_6 \div {}^2V_7$	$^1V_8$	$\{^2V_6 = {}^2V_6;\ ^2V_7 = {}^2V_7\}$	$= \frac{2n-1}{3}$	...	...	...	...	...	$2n-1$	3	$\frac{2n-1}{3}$									
16	×	$^1V_8 \times {}^3V_{11}$	$^4V_{11}$	$\{^1V_8 = {}^0V_8;\ ^3V_{11} = {}^4V_{11}\}$	$= \frac{2n}{2} \cdot \frac{2n-1}{3}$	...	...	...	...	...	...	...	0	...	...	$\frac{2n}{2} \cdot \frac{2n-1}{3}$						
17	−	$^2V_6 - {}^1V_1$	$^3V_6$	$\{^2V_6 = {}^3V_6;\ ^1V_1 = {}^1V_1\}$	$= 2n - 2$	1	...	...	...	...	$2n-2$											
18	+	$^1V_1 + {}^2V_7$	$^3V_7$	$\{^2V_7 = {}^3V_7;\ ^1V_1 = {}^1V_1\}$	$= 3 + 1 = 4$	1	...	...	...	...	...	4										
19	÷	$^3V_6 \div {}^3V_7$	$^1V_9$	$\{^3V_6 = {}^3V_6;\ ^3V_7 = {}^3V_7\}$	$= \frac{2n-2}{4}$	...	...	...	...	...	$2n-2$	4	...	$\frac{2n-2}{4}$	...	$\left\{\frac{2n}{2} \cdot \frac{2n-1}{3} \cdot \frac{2n-2}{3} = A_3\right\}$						
20	×	$^1V_9 \times {}^4V_{11}$	$^5V_{11}$	$\{^1V_9 = {}^0V_9;\ ^4V_{11} = {}^5V_{11}\}$	$= \frac{2n}{2} \cdot \frac{2n-1}{3} \cdot \frac{2n-2}{4} = A_3$	...	...	...	...	...	...	...	...	0								
21	×	$^1V_{22} \times {}^5V_{11}$	$^0V_{12}$	$\{^1V_{22} = {}^1V_{22};\ ^0V_{12} = {}^2V_{12}\}$	$= B_3 \cdot \frac{2n}{2} \cdot \frac{2n-1}{3} \cdot \frac{2n-2}{3} = B_3 A_3$	...	...	...	...	...	...	...	...	...	...	0	$B_3 A_3$	......	......	B_3		
22	+	$^2V_{12} + {}^2V_{13}$	$^3V_{13}$	$\{^2V_{12} = {}^0V_{12};\ ^2V_{13} = {}^3V_{13}\}$	$= A_0 + B_1 A_1 + B_3 A_3$	...	...	...	...	...	...	...	...	...	...	....	0	$\{A_0 + B_1 A_1 + B_3 A_3\}$				
23	−	$^2V_{10} - {}^1V_1$	$^3V_{10}$	$\{^2V_{10} = {}^3V_{10};\ ^1V_1 = {}^1V_1\}$	$= n - 3\ (= 1)$	1	...	...	...	...	...	...	...	...	$n-3$							
					Here follows a repetition of Operations thirteen to twenty-three.																	
24	+	$^4V_{13} + {}^0V_{24}$	$^1V_{24}$	$\{^4V_{13} = {}^0V_{13};\ ^0V_{24} = {}^1V_{24}\}$	$= B_7$	...	...	...	...	...	...	...	...	...	...	....	....	....	......	......	......	B_7
25	+	$^1V_1 + {}^1V_3$	$^1V_3$	$\{^1V_1 = {}^1V_1;\ ^1V_3 = {}^1V_3;\ ^5V_6 = {}^0V_6;\ ^5V_7 = {}^0V_7\}$	$= n + 1 = 4 + 1 = 5$; by a Variable-card. by a Variable card.	1	...	$n+1$	...	...	0	0										

Written by Ada Lovelace in 1842, this is the first algorithm ever published.

What is assembly language?

Assembly language (or assembler language) is a low-level program that is designed for a specific type of computer architecture. Sometimes, programmers will write a code for a specific process in assembly language to make sure it functions as efficiently as possible.

Who was Charles Babbage?

Charles Babbage (1791–1871) was a professor of mathematics at Cambridge in London. He is known as the creator of the difference engine and the analytical engine. He is known as the Father of Computing. During his youth, Babbage taught himself algebra, and in his 20s, he was consumed with ideas about machines that could perform calculations. The first machine that he developed was the "difference engine," which could only perform one task at a time. His second machine, the "analytical engine," was designed to perform different types of mathematical tasks. Unfortunately, Babbage only had prototypes of these machines and never completely manufactured them. He worked with Ada Lovelace to document the designs and describe their functions. After his death, his machine was constructed using the blueprints and articles that Lovelace created, and both the British and American governments used the machine.

The Father of Computing, Charles Babbage, shown here in 1860.

What does a programmer have to consider when designing a new program?

Programming a computer is similar to developing a recipe that must be followed. The programmer must know what instructions must be followed, the order that those steps must be followed, and what tools are needed to follow the instructions. They might think of the program itself and the inputs (tools), process (program), and outputs (goal) of the project.

What process does a programmer use to write code?

Writing code can be done in many different ways. Each programmer will develop their own way. One method that a programmer can use is called the program development cycle. It involves nine phases. Even before beginning the process, the programmer has to identify the problem. Then, the program development cycle begins with problem analysis. In the next step, the program developer generates multiple solutions to the problem. The programmer identifies from that set of solutions the best solution and presents the solution by writing a sequence of steps. This sequence of steps, called an algorithm, is tested by "walking through" each step to check for errors before converting the steps to a code. Once the code is written, it must be tested. The program developer documents the details of the program and then installs the program, and the user begins to operate and understand the program. Once the program is implemented, the programmer will regularly check the program for updates.

What is a programming language?

Programming language is what allows humans and computers to interact. The computer will not perform tasks that it was not programmed to do. Some of the top computer programs are:

- JavaScript
- SQL
- Java
- Python
- C#
- PHP

- C++
- C
- TypeScript
- Ruby

What is the Python programming language?

Python was created in the 1980s by a man named Guido van Rossum (1956–). Currently, it is the most popular and most used programming language. It is considered a high-level programming language. This means that it is not a machine language. It has elements that look like the English language and needs an interpreter to compile it so that the computer can understand the instructions the user is giving it.

Has the Python language changed since it was created in the 1980s?

Yes! Over time, programming languages evolve and are updated as more people begin to use them and improve them. It is now called Python 3.

Where is the Python language used?

The Netflix streaming service uses the Python language to analyze user data and for security. Some video games, such as *Battlefield 2*, use Python for

Dutch programmer Guido van Rossum invented the Python programming language.

add-ons and functionality. Government agencies use Python for secret purposes. Google uses Python because it allows the company to build programs easily.

What are the benefits of Python?

Using the Python language can benefit the programmer because the language has many existing codes called libraries that have common program features. For example, if a programmer needs a code for a very complex mathematical equation, they can check the library to see if one is already written, which will save them lots of time. Many programmers find Python easier to read and diagnose quickly. Python also runs on many systems and is very flexible. It can be used for many different purposes and in many different job functions.

What is the C programming language?

The C language was first introduced in 1972 by Dennis Ritchie and since then has been one of the most used mid- to high-level programming languages. Some of its features provide the programmer with direct access to the computer's memory and other basic computer functions. This aspect of the language can make it difficult for some new programmers to learn. This language still needs an interpreter for the computer to understand the instructions.

Where is the C language used?

The C language has some key applications. Software developers use the C language in embedded systems and for developing computer desktop applications. Developers use the C language when they are developing computer operating systems; for example, the Apple operating system OS X, the Microsoft Windows operating system, and even operating systems found on cell phones. Google Chrome is a web browser that was built using the C language.

What is an example of a program that is written in the C language?

Here is a program to display "Hello, World" on a screen:

DID YOU KNOW?

What are the benefits of the C language?

Some features of the program, such as having direct access to the computer's memory, create opportunities for the programmer to learn how computers work! Since C is the base language for many other languages, learning C will help a programmer learn and understand other languages. The C language has a large library, and users can use code from the library or add their own code to the library.

```
#include<tdio.h>
int main()
{
printf("Hello, World");
return 0;
}
Output:
Hello, World
```

What does this program mean?

#includeLESS THANstdio.hGREATER THAN

This is a command that tells the compiler to include the functions of the stdio.h header file in the current program.

int main()

The function name is main(), and the return type of this function is int. Every C program has this function.

{ and }

The C language is a block-structured language, so statements in the program are grouped together using these symbols.

printf("Hello, World");

printf is a function in the stdio.h header file. This function is used to print the phrase "Hello, World" on the screen.

return 0;

DID YOU KNOW?

Can computers understand English?

Computers actually only understand one language, which is called "machine language." The computer recognizes two states: "on" or "off." These states are represented by 0s and 1s. Computers do not understand English or any other higher-level language, so it must be translated from programming language to machine language by a compiler, interpreter, or assembler. The type of translator used depends on the programming language that needs to be executed.

When each line of the program is executed without errors, it will get to this line. If it contains errors, then the program returns an error.

How has the C language evolved?

The C language forms a base for many other languages, including C++ and Java. At one point, so many different versions of the language were being used that it was hard for developers to keep up with the latest version. Eventually, an organization called the American National Standards Institute (ANSI) created and approved a standard version.

What is a programming language used for?

A programming language is used to provide instructions to a computer in a language that it understands.

What is debugging?

Debugging is the process of reviewing lines of code for errors. Errors in code are also called bugs. The two types of bugs are logic bugs and syntax bugs. With a logic bug, the program will run, but the computer does not do what the programmer intended for it to do. With a syntax bug, the computer will not be able to run the program because it does not understand what was typed.

DID YOU KNOW?

What is artificial intelligence (AI)?

Artificial intelligence is the ability of machines to learn and think in a similar manner to humans. An American scientist named John McCarthy came up with the term in 1955. Currently, AI is used to recognize speech, as bots who answer questions on websites, and to drive cars, among other uses. In fact, to teach machines how to make intelligent decisions, researchers train computers to train and beat old video games! The computers use trial and error, store the trials in memory, and then recall the different scenarios to learn from mistakes and win the games.

What is a strategy that can be used to understand code?

Decomposition is a problem-solving strategy to break down a large problem into smaller problems that are easier to understand and solve. In computer science, rather than making a large and complex program, the programmer would break up the large programs into smaller chunks of code. Then, when an error occurs in the code, the programmer does not have to review many lines of code at one time. They can review each chunk of code in separate passes.

THE INTERNET AND NETWORKING

What does WWW stand for, and what is it?

The World Wide Web is a collection of all the websites from all over the world that are found on the internet. Each computer uses a web browser like Google Chrome or Firefox to access the web. It is a way to share information with people from all over the world very quickly.

How is the World Wide Web different from the internet?

Many people think that the internet and the World Wide Web are the same. The internet is the global network of all the computers connected together. Millions of computers are connected through cable, fiber-optic, and wireless links. Each computer has an address called an IP (Internet Protocol) address.

Who founded the World Wide Web and when?

The World Wide Web was created in 1989 by Sir Tim Berners-Lee. He was a British scientist who wanted to create a way for scientists, universities, and institutes to share files. Berners-Lee did not launch the first website until 1991. The first websites were simply words and pictures. It wasn't until 1993 that the Web was available for free to the world. At that time, websites were made by people learning and writing HyperText Markup Language (HTML) code.

Here is a link to the very first Web pages and other interesting links related to the history of the World Wide Web: http://info.cern.ch/

What is a computer network?

The foundations of a computer network are switches, routers, and wireless access points. Each of these devices serves an essential role in connecting all of us to the internet and providing access to the Web. Every device that is connected to the internet is a part of a computer network.

Credit for the important invention of the World Wide Web can be attributed to English computer scientist Sir Tim Berners-Lee.

Computer networks are made up of the following elements:

- Servers and clients are the two categories that all internet machines fall into. A machine provides a service to other machines; then, it is called a server.
- Switches connect devices such as computers, printers, and servers. This creates a network of resources. Switches are very important in business, office building, and campus settings because they contain many computers, printers, and other devices that must be connected to each other. One way to illustrate how a switch works is to view it as a controller.
- Routers connect multiple networks of resources to each other and to the internet (global network of devices). Routers can save a company lots of money because they allow all the networked devices to share a single internet connection. Routers are also used at home to share a connection to the Web with all the devices connected to it, behaving as a hub.
- Modems connect devices to the internet. Each home has an internet service provider (ISP), which provides access to the internet through fiber-optic cable, coaxial cable, or copper (DSL) lines. The modem connects the cable from the ISP and translates the cable signal into a language the computer can understand. A computer can have a direct physical connection to the modem through an ethernet cable.
- Access points allow multiple devices across multiple networks to connect to each other without cables. The access point acts as an amplifier for the network. It is another way for devices to connect to the network.
- Firewalls are the first line of defense against cyberattacks.
- Layered security protects all of the local devices, switches, and routers.

CAREERS

What skills are needed to become a computer engineer or computer information scientist?

All computer engineers and computer scientists begin their education with computer programming. In addition to programming, a computer scientist/computer engineer needs to be able to solve complex problems, and they have to invest in learning for the rest of their careers.

- Computers and electronics
- Engineering and technology

- Design
- Mathematics
- Physics
- Production and processing
- Telecommunications

What types of work do CEs/CISs do?

Computer engineers work in very diverse places, including in aerospace, power, manufacturing, defense, and, of course, with computers and electronics. They design microelectronic chips and the powerful systems that use those chips. A telecommunication system is the system that allows people to communicate from a distance using the computer software and hardware that is designed by computer engineers.

What are some common problems that CISs/CEs solve?

The work of computer scientists and engineers can be seen around us in our everyday lives. Some of the most common problems their work helps us to solve can range from being able to communicate with others for social or business purposes, having access to information, or just simply making our lives easier.

In terms of communication, it is through the invention of and constant maintenance of communications systems such as satellites, cell phones, computers, and televisions that we are able to keep in touch with friends and family and engage in business- or school-related communication. In this instance, computer scientists and engineers must keep abreast of emerging technologies and how these technologies may impact our current infrastructure. When changes in technology affect the efficiency of our devices or platforms, these experts design software updates and new hardware elements that are then used to ensure the continued smooth functioning of our devices.

Computer scientists and engineers are also always involved in ensuring that we have access to information. All of our devices that we use to gather or access information are designed by computer scientists and engineers. People use the internet every second of the day to find out about current events, information about their work or hobbies, or just to learn new things. The work of computer scientists and engineers is very important in this regard.

Name an electronic convenience you enjoy in your home or office or school today, and you can thank engineers and computer scientists for most of them.

Computer scientists and engineers influence every aspect of our lives and use their knowledge and expertise to make our lives easier. Almost everything in our immediate environment is influenced by computers. Ensuring that our televisions, cell phones, computers, refrigerators, washing machines, and even cars work as they are expected to are some of the common everyday problems that computer scientists and engineers solve on a daily basis.

What are the top 10 contributions that CISs/CEs have made to society?

Computer scientists and engineers have designed and developed several inventions that have been of great benefit to society. Some of these contributions are:

1. Being able to communicate with others across the world.
2. Making research easier.
3. Increasing the ability to buy and sell goods, thereby increasing the quality and quantity of goods we have access to.
4. Increasing our access to information, thereby enhancing our knowledge.
5. Improving health care.
6. Advancing medicine through improved research capabilities.
7. Being able to tend to school and work tasks in a virtual environment, thereby increasing how many collaborators we have.
8. Increasing entertainment options.
9. Being able to perform everyday tasks from anywhere in the world, such as paying bills, online shopping and banking, etc.

10. Helping to create a better world by giving access to various cultures, societies, and ethnic and racial groups that may not otherwise have been able to participate in various spaces.

What is the future of CISs/CEs?

Computer science is a quickly developing field with a steady increase in demand for computer science graduates in a wide variety of fields. A few areas that will be trending in the years to come include artificial intelligence and robotics, big-data analytics, computer-assisted education, bioinformatics, and cybersecurity.

Aerospace Engineering

What is aerospace engineering?

Aerospace engineering is the primary field of engineering concerned with the design, development, testing, and production of aircraft, spacecraft, and related systems and equipment. Aerospace engineers design primarily aircraft, spacecraft, satellites, and missiles. In addition, they test prototypes to make sure that they function according to the design of aircraft, spacecraft, missiles, and other airborne objects. Creating prototypes of these designs and testing them is also a primary responsibility of an aerospace engineer. But aerospace engineers also evaluate designs to see that the products meet engineering principles. Simply put, aerospace engineering is the process of designing, developing, testing, manufacturing, and producing anything that flies.

What are the different branches of aerospace engineering?

Aerospace engineering has four main branches. This chapter will answer many questions related to the design of flight vehicles. This requires knowledge from many engineering disciplines, and some individuals and companies specialize in addressing these complex design challenges.

- Aeronautical engineering involves the science and technology of aircraft design and propulsion systems inside of Earth's atmosphere. In recent times, aeronautical engineers have commonly worked on military, commercial, and passenger jets; helicopters; and drones. In this field, an engineer can plan, design, and conduct wind tunnel tests for various aircraft models or different aircraft components, design and analyze airplane wing structures, and design and test jet engines. Aeronautical engineers use theory and technology and understand the fundamentals of flight to study the aerodynamic performance of materials.

- Astronautical engineering involves work with science and technology on aircraft or spacecraft that operate outside of Earth's atmosphere. This requires knowledge of space and the complexities of operating technology in space. Often known as "rocket science," in this field, an engineer can design and test rocket engines; design, analyze, and optimize orbital trajectories (space routes); and develop control systems for rocket propulsion and space flight.

- Aviation engineering has to do with the flying, developing, and maintenance of an aircraft and the systems on board. It is the practical side of aeronautical science. This means that an aviation engineer's work is concerned with designing, developing, maintaining, troubleshooting, upgrading, and assembling aircraft. These engineers might perform traditional testing and design with actual planes, but they are also skilled in software applications that allow them to simulate testing environments and flying conditions in order to evaluate the performance of their designs. Because their work often addresses issues related to efficiency, aircraft noise, costs of production and operation, or even reducing air emissions, aviation engineers work well with other engi-

Aerospace engineers can be found working for the government and private sectors on everything from airplanes to drones, spacecraft, satellites, and missiles.

DID YOU KNOW?

What types of problems do aerospace engineers solve?

Most problems in aerospace relate to aerodynamics, propulsion, performance, stability, control, structures, and systems.

neers and professionals to solve problems. They work as part of a team and often work on a specific part of the aircraft. These folks are sometimes called technicians because their work does not require an engineering degree, but as one can imagine, it does require a lot of treatment and education.

- Avionics engineering involves the electrical and electronic systems of an aircraft or spacecraft. The word avionics is derived from the words "aviation" and "electronics." Avionics engineering has a specific focus on the electronics applications and challenges of aircraft, spacecraft, and missile design and operations. An avionics engineer might work on projects related to designing instrument launching and landing systems or safety instruments. For example, Hibah Rahmani, an avionics and flight controls engineer at NASA, supports the organization's launch services program and focuses on launch vehicle testing. In order to study avionics engineering, some core courses include electronics, aviation systems, communications, microwave, radar, antenna, guidance, navigation, and controls. Some of the aviation electronics that people are familiar with include navigation, weather radar, flight recorders, and fuel systems.

What do aerospace engineers do?

Aerospace engineers primarily design, analyze, research, develop, and test aircraft, spacecraft, satellites, missiles, and other technology. In addition, they create and test prototypes to make sure that they function according to design. They rarely work alone. They work to ensure that the vehicle—whatever it is—behaves in flight as they predicted it would. Typically, aerospace engineers work on design teams to develop solutions to complex vehicle design problems. Some of the roles an aerospace engineer might fill on a design team include specializations in:

DID YOU KNOW?

What is the difference between an aeronautical engineer and an astronautical engineer?

An aeronautical engineer works on planes and their parts, whereas an astronautical engineer focuses on spacecraft and related technologies. The air in the atmosphere provides support for aircraft and other atmospheric flying vehicles. This is accomplished by using aerodynamics to counteract gravity. Since spaceships do not rely on air or atmosphere for lift or propulsion, they do not require this aerodynamic design.

- Aerodynamics science
- Propulsion systems
- Structural design
- Materials
- Avionics
- Stability and control systems

How is aerospace engineering different from aerospace science?

Aerospace engineering differs from aerospace science in that engineering is more related to the practical application of space science principles. Aerospace scientists are better known as astronomers and astrophysicists. Astronomers are natural scientists who study space, stars, planets, meteors, and galaxies. They share some equipment with aerospace engineers—for example, radio and optical telescopes like the infamous Hubble Space Telescope. They expand our scientific knowledge of space. That knowledge can be used by engineers to develop new technologies related to communication and energy storage and the medical field to make breakthroughs on new medicines. Scientists expand fundamental science knowledge and then engineers draw upon that knowledge, apply it, and develop and improve innovations.

HISTORY

Who is the founder of aerospace engineering? What are its roots?

Aerospace engineering has its roots in mechanical engineering and the study of aerodynamics. Aerodynamics is a branch of physics. The history of aerospace engineering includes more than airplanes and spacecraft. Historic flight events included the flights of objects such as hot-air balloons, helicopters, and artificial satellites. Aerospace engineering as a term first emerged in 1958 and considered Earth and space as one single realm of study. But recently, aerospace engineering now includes two main branches, aeronautical and astronautical engineering, that encompass both Earth's atmosphere and outer space, respectively.

Many innovators, designers, and enthusiasts developed air vehicle concepts in the 1500s to the 1900s. This timeline will highlight some major milestones.

Who are some famous aerospace engineers?

- Italian Leonardo da Vinci (1452–1519) sketched flight vehicles and proposed flight ideas (1400s).
- French brothers Joseph-Michel (1740–1810) and Jacques-Étienne Montgolfier (1745–1799) designed a hot-air balloon, which flew over Paris as the first manned flight (1783). The pilots burned wool and straw to keep it afloat.
- George Cayley (1773–1857): One of the most important people in the history of aeronautics and regarded as the Father of Aviation. An English engineer, inventor, and aviator, Cayley is viewed as the first scientific aerial investigator and was one of the first people to understand the principles of flight.
- French inventor Henri Giffard (1825–1882) invented the dirigible balloon, or airship, which included a self-propulsion system (dirigible) that allowed for forward motion. It included a cigar-shaped bag (balloon) filled with gas that was lighter than air, a holding place for the crew that hung beneath the balloon, engines to drive the propellers, and rudders for steering (1852).
- German scientist Otto Lilienthal (1848–1896), among other individuals, began to collect flight data in order to study aerodynamics and aircraft design by flying thousands of flights (1891).

- American brothers Orville (1871–1948) and Wilbur Wright (1867–1912), the fathers of modern manned flight, designed and flew the first heavier-than-air vehicle in 1903. They eventually sold their model to the U.S. Army.
- Robert H. Goddard (1882–1945): The American Father of Rocketry. He built and tested the world's first rocket in 1926. He advanced theory about rocket design and made practical rocket design advancements. His rocket studies first began with gun power experiments.
- Barnes Wallis (1887–1979): Most famous for his invention of the bouncing bomb, used by the Royal Air Force during the "dambusters raid" during World War II. He was an English scientist, engineer, and inventor.
- Reginald Joseph Mitchell (1895–1937): An English aeronautical engineer whose most famous aircraft design was that of the iconic Supermarine Spitfire, the World War II fighter.
- Sergei Korolev (1907–1966): Trained in Russia in aeronautical engineering and developed Russia's first rocket-propelled aircraft. However, during a political fight, he and other scientists and engineers were thrown in prison. The country realized that they needed aeronautical engineers for war advancements and created special prison-based design bureaus to use their talent and knowledge. Korolev worked on these special bureaus and was later released from prison. His effort to help Russia win the Space Race against the United States was cut short with his death in 1966.
- Hans von Ohain (1911–1998): A German physicist who designed the first operational jet engine, which was used to power the world's first flyable all-jet aircraft in 1937.

Orville (left) and Wilbur Wright were the first to successfully fly a heavier-than-air vehicle (airplane), which they did at Kitty Hawk, North Carolina, on December 17, 1903.

- Wernher von Braun (1912–1977): A German-born American aerospace engineer and space architect who was called the Father of Rocket Technology and Space Science. He was the chief architect and director of the Marshall Space Flight Center, and his work later proved to be instrumental in propelling the Apollo mission to the moon.
- Neil Armstrong (1930–2012): The first American astronaut and aeronautical engineer to walk on the moon. He was known for the saying, "That's one small step for man, one giant leap for mankind."
- Colonel Guion Stewart Bluford Jr. (1942–2003): A NASA astronaut and the first African American to fly in space. Bluford was a retired U.S. Air Force officer and fighter pilot. His first flight was the STS-8 aboard Challenger in 1983 and his last was in 1992.
- Dr. Kalpana Chawla (1962–2003): The first India-born woman in space. She earned her first engineering degree in her home country and later moved to the United States, where she became a naturalized citizen. She began working for NASA in 1988, and by 1994, she was an astronaut candidate. She traveled in space in 1997 and again in 2003. Unfortunately, after completing many experiments in space, she and her space crew were killed when Columbia disintegrated upon reentry. She is famous for saying, "When you look at the stars and the galaxy, you feel that you are not just from any particular piece of land but from the solar system."
- Elon Musk (1971–-): An South African entrepreneur, engineer, and the CEO of SpaceX, an aerospace company that manufactures spacecraft and is designing a space vehicle that would make space travel more accessible for ordinary people. He is also trying to develop a way to establish a human colony on Mars.

How do hot-air balloons work?

Hot-air balloons, as the name suggests, are based on the basic concept that warmer air rises in cooler air. Thus, the underlying scientific theory behind hot-air balloons is as follows:

- Since heated air is lighter than cool air, it has less mass per unit of volume. The weight of a cubic foot of air is approximately 1 ounce, but once we raise the temperature of the air by 100 degrees Fahrenheit, a loss of 7 grams of weight occurs.
- As a result, each cubic foot of air in a hot-air balloon can lift around 7 grams.
- For example: In order to carry 70 stones, we'll need roughly 65,000 cubic feet of hot air; this is why hot-air balloons are so massive.

Brothers Joseph-Michel (left) and Jacques-Étienne Montgolfier created the balloon in France that was the first to have a manned flight over Paris.

How did aircraft designs change throughout history?

Early aircraft designs were influenced by nature. For example, designers observed the way that birds fly and incorporated aspects of bird features into aircraft. The first controlled, sustained flight occurred in 1903. Orville and Wilbur Wright were brothers who spent time reviewing existing technical information on theory and design. From 1900 to 1903, the brothers performed many thorough flight tests with gliders, both manned and unmanned. The Wright brothers' first powered airplane was called the *Flyer I*; it had a maximum air speed of 30 miles per hour and weighed 750 pounds.

Who made the earliest sketches of flight vehicles?

Leonardo da Vinci drew the first sketches of flying vehicles. The first image was of a machine with flapping wings like a bird called the ornithopter. The second was similar to today's helicopter.

Was Leonardo da Vinci an aerospace engineer?

No; Leonardo da Vinci was not an aerospace engineer, but he was responsible for the earliest sketches of flight vehicles. He was a brilliant, fifteenth-cen-

When we think of great designers of old times, the man who springs to mind is often Leonardo da Vinci (1452–1519), who created ideas for such incredible devices as steam-powered cannons and an early version of the helicopter.

tury Italian painter, sculptor, architect, and engineer who was responsible for painting popular works of art, including the *Mona Lisa* and *The Last Supper*. Mechanical motion and machines were very interesting to him, and he would engage in research to understand the forces that made nature move. For example, he studied the flight of birds and bats, which influenced his design of the ornithopter. His conceptual drawings of flight vehicles included the parachute, the helicopter ("aerial screw"), a fighting vehicle, and other man-powered flying machines. Many of his drawings could function exactly as drawn with minor modifications and additions of mechanical elements.

When was the first successful flight?

Early innovators of powered, lighter-than-air flying vehicles included Jules Henri Giffard, who in 1852 flew the first steerable, steam-powered airship; Charles Renard and Arthur Constantin Krebs, who in 1884 flew the first powered airship that returned to its starting point; and Ferdinand von Zeppelin, who built and flew the first rigid airship in 1900. On December 17, 1903, Wilbur and Orville Wright made four brief flights at Kitty Hawk, North Carolina, with their first powered aircraft. The Wright brothers had invented the first successful airplane. Robert H. Goddard, an American, developed, built, and flew the first successful liquid-propellant rocket on March 16, 1926.

AIRPLANES

What are the parts of an airplane?

An aircraft is made up of many individual parts. The typical parts of an airplane are:

Windshield	Cockpit
Fuselage	Engine
Wheel	Strut
Antennae	Slat
Flap	Spoiler
Wint	Winglet
Aileron	Vertical stabilizer
Rudder	Horizontal stabilizer
Elevator	Empennage

What are the four forces that help airplanes fly?

The four forces are as follows:

- Thrust is the force that propels an aircraft forward in its motion. A propeller, jet engine, or rocket is used to make it. Air is drawn in one way and then pushed out the other.
- Drag is the force that operates in the opposite direction as motion. It has the effect of slowing down an item. Friction and variations in air pressure generate drag.
- Weight is the force of gravity.
- Lift is the force that keeps an airplane in the air. The wings provide the majority of the lift that aircraft require.

The plane does different things depending on how the four forces interact with each other. Each force has a counterforce that opposes it. Lift operates in the opposite direction of weight. Drag is the polar opposite of thrust. A plane flies in a level direction when all the forces are balanced; for example, if the forces of lift and thrust are greater than weight and drag, the plane will rise.

DID YOU KNOW?

What is the U.S. Federal Aviation Administration?

The U.S. Federal Aviation Administration (FAA) is the federal agency that oversees the aviation operations and safety in the United States. The agency establishes and monitors FAA regulations and policies, oversees the aircraft certification process, and controls the licenses and certifications of those who operate and perform maintenance on aircraft. The Civil Aviation Authority oversees and regulates civil aviation in the United Kingdom.

What are flight dynamics and control?

Flight dynamics and control deal with the motion of flight vehicles both in and outside of Earth's atmosphere. The four basic flight maneuvers are straight and level, turn, climb, and descent.

Who makes sure that planes and flight vehicles are safe to fly in?

In the United States, the Federal Aviation Administration (FAA) is in charge of air safety.

What are some important aerospace engineering terms?

- Aircraft is a term that includes all types of vehicles that fly within Earth's atmosphere. They are classified as either lighter than air or heavier than air. Lighter-than-air aircraft support their weight differently than heavier-than-air aircraft. A balloon is a commonly recognized lighter-than-air aircraft. Heavier-than-air aircraft include gliders, airplanes, and helicopters.
- A spacecraft is a vehicle that engineers have designed to travel through space. The first spacecraft was designed by a very large team and

launched into space on October 4, 1957. It was unmanned and piloted by Yuri Gagarin. Spacecraft are powered by satellites and missiles.

- Aerodynamics is simply the way air moves around objects, the science of the flow of air and its interaction with bodies in the air. Bodies might include airplane or animal wings, automobiles, buildings, or airplanes. Two specific concepts at the center of aerodynamics are lift and drag. Lift is the force that opposes the weight of the bodies in the air and is the force that keeps the body in the air. Lift is produced as an airplane (object) moves through the air. Lift occurs when the direction of air is turned by an object (i.e., airplane) moving through it. If air is not present, then no lift can occur. If the airplane (or other object) is not in motion, then no lift can be generated. Drag is a force that is also generated by an airplane (or other object) moving through the air. Drag resists the motion of an airplane moving in the air. Engineers whose focus is aerodynamics create designs and shapes that create a lot of lift with little drag.

- Force is a push or pull in a specific direction. Whenever objects interact, a force is placed upon each object. Force is a result of that interaction. Force has a size that can be measured (magnitude) and a direction (i.e., upward, downward, forward, or backward).

- The four aerodynamics forces—lift, drag, weight, and thrust—are the four forces of flight that act on an airplane. The forces make objects move in different directions at different speeds. Basically, force determines how the object moves through the air. The movement of the airplane in the air depends on the magnitude and direction of each of these forces on all parts of the plane. If all of the forces are balanced, then the plane cruises at a steady velocity, but if the forces are unbalanced, then the airplane accelerates in the direction of the force that is the strongest: the greatest magnitude.

- The weight of an object is the force on the object caused by gravity. Since gravity affects everything on Earth, then everything on Earth has weight. It might be easy for humans to understand weight because they are typically very familiar with their own weight. Human weight is measured using scales. For an airplane, the weight has magnitude and direction and is directed toward the aircraft's center of gravity. The magnitude is based on the mass of every single object in the plane and all the parts that make up the plane. Remember that modern planes use gas to travel; therefore, the weight changes, and so does the center of gravity. The pilot has to pay attention to all of these changes in order to keep the plane balanced during flight.

- Lift is one of the major forces acting over a plane in flight and is the opposing aerodynamic force that counters the weight force. This is the force that overcomes gravity and holds the object in the air. The wings of the aircraft generate most of the lift. Lift is generated when a moving airflow interacts with an object. The air moves in one direction, and the object (aircraft) moves in another. Lift acts through the center of pressure. As the air moves around the aircraft, the speed of the air produces various pressures all over the aircraft. This means that the vehicle speed, size, and shape all affect the lift. The center of pressure is the one point through which all the lift is focused. As the aircraft changes direction and velocity, the center of pressure will also change. When the flight is balanced, then weight = lift. Here are a couple of fun rules to remember about lift: no fluid (air), no lift; no motion, no lift.

- Drag is the aerodynamic force that resists the motion of the aircraft through the air. It is generated by the difference in velocity between the aircraft and the air. If no motion occurs, no drag can occur; therefore, if the plane is on the ground, no drag occurs. A simple representation of drag is friction. When the aircraft surface is smooth, waxed, and sleek, less friction (drag) occurs, but when the surface is rough, then more friction (drag) occurs. Consider trying to slide across a polished, wooden floor while wearing socks and comparing it to doing the same on a floor of pebbles. In addition to the surface quality, the shape of the aircraft also affects drag. As pressure changes around the bodies, then the forces acting on the aircraft also change. Drag is another force that acts through the center of pressure of the aircraft. In steady cruising aircraft travel, flight thrust = drag.
- Thrust is the force that moves the aircraft through the air. Thrust is generated by the engines, which bring air on board, mix the air with fuel, and then burn the mixture. The fuel exhaust produces the thrust needed to move the aircraft through the air. Thrust overcomes drag and the weight of the aircraft. The engines are part of the propulsion system, and they must interact with moving air in order to make the aircraft

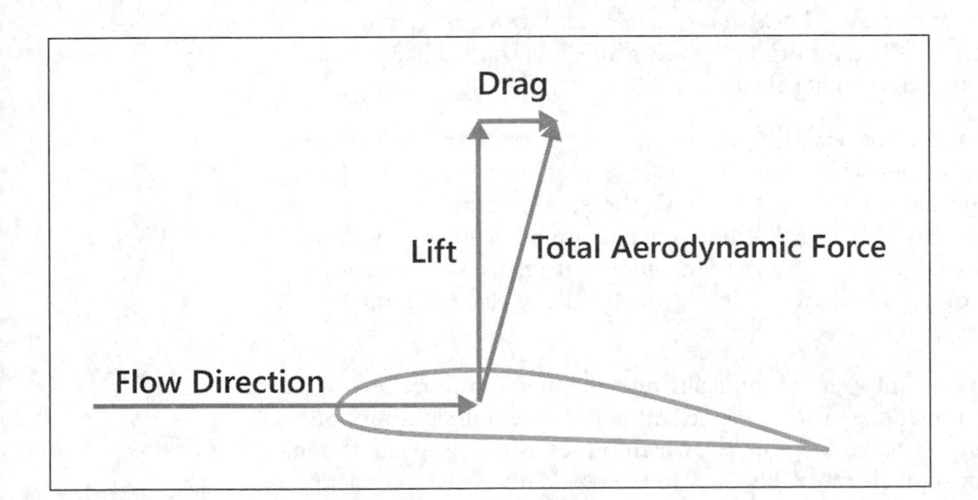

Figure 17: A diagram showing lift and drag affecting the vectors as they act on an airfoil.

move. Direction of thrust opposes the direction of airflow. The engines pull air through the system, and the aircraft moves forward. Thrust must be equal to drag for an aircraft to fly.

- Turbulence is the movement of air caused by some disturbance like atmospheric pressure, jet streams, cold or warm weather fronts, or thunderstorms. The four main types of turbulence are mechanical, thermal, shear, and aerodynamic.

- An airfoil (or aerofoil) is any surface that is designed to create lift and drag when in motion. An airplane wing, blade, or sail can act as an airfoil. Other airfoil surfaces include tailplanes, fins, winglets, and helicopter rotor blades. Multiple airfoil sections can be on one aircraft part. This means that small forces are acting on various sections of the airplane. All of the small forces together equals the lift. The small forces acting on various sections of the aircraft part (lift) can be used to determine the center of pressure of an aircraft. Aircraft designed to have reached high speeds have streamlines and thin airfoils that have low drag and low lift. Aircraft that carry heavy loads are typically slower, and they use thicker airfoils that have high drag and high lift.

- Flow fields describe the motion of air around an object. They are determined by the airflow velocity, pressure, density, and temperature.

- Propellers are objects that convert the spinning motion of the engine into a forward motion. Propeller blades are twisted and angled to make sure that a constant thrust is occurring along the length of the blades. This is because different parts of a propeller move at different speeds. The tips (outermost part of blades) move faster than the section of the blade that is closest to the hub of the propeller. Aircraft propellers are designed to move fast in air, so they have a thick and narrow blade design. Materials used to make propeller blades include lightweight aluminum or magnesium alloys, hollow steel, and wooden laminates or composites. The Wright brothers used propellers shaped like airfoils in their first powered flight vehicle in 1903.

- Aircraft engines mix air with gas to propel the aircraft forward; the propellant is the chemical mixture that is burned with the air that is pulled through the engines in order to produce thrust or motion. The propellant includes fuel (something that burns when combined with oxygen and produces gas) and an oxidizer (an agent that releases oxygen for the purposes of combining with fuel). Propellants are classified into states: hybrid, gas, liquid, or solid.

- A solid-propellant system is typically referred to as a motor. One of the main components of a solid propellant is the case. Other components include the igniter and the nozzle, which serves as the propellant's tank and combustion chamber. The fuel and oxidizer are mixed together and cast into a solid mass through a polymerization process. The molds are removed after the solidifying process.

- The main components of a liquid-propellant engine are the combustion chamber, fuel tank, oxidizer tank, injector, and supersonic nozzle. The liquid fuel and liquid oxidizer are stored separately and at a pressure above that of the operating pressure of the engine. The propellants are injected into the engine in a way that ensures atomization and rapid mixing. Liquid propellants are preferable to and have more benefits than other types of propellants.

What is propulsion?

Propulsion is the forward motion of an object. The four main propulsion systems are the rocket, turbine or jet, propeller, and ramjet. In an airplane, the forward motion or forward propulsion can be generated by the engine. The engine helps the airplane (aircraft) balance drag in order to cruise at a constant velocity, and an engine also helps the plane exceed drag to accelerate. Aeronautical engineers can design different types of engines that can produce different amounts of thrust based on the power and thrust needs of the aircraft.

What is a propulsion system?

The purpose of a propulsion system is to produce movement. The most common type of rocket engine is a chemical type where hot exhaust gases are produced by chemical combustion.

What have been some of the historical challenges to aerospace engineering?

Aerospace engineering emerged from mechanical engineering. During this time, researchers and scientists had a growing interest in understanding flight vehicles and innovative concepts related to that. To date, the earliest sketches available of flight vehicles were drawn by Leonardo da Vinci. He is most notable for his very detailed drawings. He suggested multiple designs to keep what we call a plane in the air. That was the first historical challenge: what design would keep the body of a vehicle in the air.

An additional challenge was how flight could be used to move people from place to place. This is called a "manned flight." The first manned flight occurred in 1783. However, it was not in a plane like what we see today. The first manned flight actually occurred in a forward-moving hot-air balloon designed by brothers Joseph-Michel and Jacques-Étienne Montgolfier.

What is an Air Traffic Control (ATC) clearance? When is it permissible for a pilot to deviate from an ATC clearance?

An ATC clearance is permission granted by the ATC to an aircraft to operate inside controlled airspace under defined circumstances in order to avoid collisions between known aircraft. A pilot is not authorized to violate any minimum-altitude condition or regulation, deviate from any rule, or to operate the aircraft in an unsafe manner.

It is permissible for a pilot to deviate from an ATC clearance only under all of the following conditions:

- An updated clearance is acquired
- An emergency arises
- The deviation is in accordance with a traffic warning and collision-avoidance system resolution recommendation
- The operation is being performed in Visual Flight Rules (VFR) weather circumstances (except in Class A airspace, which is 0 to 4,000 feet above airport elevation)

Avionics

What is avionics?

Avionics are the electronics or electronic equipment and systems in an aircraft.

What is supersonic speed?

Supersonic speed is the rate of travel that exceeds the speed of sound.

How is supersonic speed used by aerospace engineers?

Spacecraft fly at supersonic speed. Aerospace engineers study supersonic speed to develop new space shuttles that can perform well at supersonic speed. They study supersonic flight using model spacecraft in wind tunnel demonstrations.

What are the three major subdivisions in electronic warfare?

Electronic warfare is divided into three categories: electronic attack, electronic protection, and electronic warfare support.

- Electronic attack: This includes the use of electromagnetic energy, directed energy, or antiradiation weapons to strike troops, facilities, or equipment with the goal of weakening, neutralizing, or eliminating enemy combat capabilities.
- Electronic protection: This includes both passive and active measures used to safeguard people, facilities, and equipment from the impacts of friendly or hostile electronic warfare operations that degrade, neutralize, or eliminate friendly combat capabilities.
- Electronic warfare support: This entails actions directed by an operational commander to intercept, identify, and locate sources of radiated electromagnetic energy for the specific purpose of immediate threat recognition, targeting, planning, and execution of future operations.

Part of the electronic protection systems in electronic warfare, this U.S. Navy E2C Hawkeye is equipped with early warning and command systems.

SPACE AND SPACECRAFT

What were the first successful space missions?

The Soviet Union launched the first satellite into space, Sputnik 1, on October 4, 1957. It orbited Earth for two months before falling back to the planet on January 4, 1958. The satellite carried only primitive radio equipment and broadcasted a signal heard all over the world by government, professional, and amateur radio enthusiasts. Panic ensued in the United States over the launch, and this sent us into the Space Race.

It was also a Soviet pilot (cosmonaut), Yuri Alekseyevich Gagarin (1934–1968), who became the first man in space. Gagarin completed one orbit of our planet aboard the Vostok 1 on April 12, 1961.

When was the first successful U.S. flight into space?

The Explorer 1 was the first U.S. satellite to orbit on January 31, 1958. Alan Shepard was the first American to leave Earth's atmosphere and make it into space in 1961. John Glenn was the first American to orbit Earth on February 20, 1962.

Soviet cosmonaut Yuri Gagarin was a national hero when he became the first man in space in 1961. Sadly, he died in 1968 during a training exercise when his MiG-15UTI crashed.

DID YOU KNOW?

What was the Space Race?

The Space Race was ignited after World War II between the United States and what was then the Soviet Union to see which country would be the first to develop capabilities for space travel and exploration.

What are rocket engines?

A rocket engine is a reaction engine that propels a vehicle using the thrust generated by the high-temperature gas that is expelled through a nozzle, which causes the vehicle to travel at high speeds. The reaction may be from a chemical or nuclear reactor.

Who developed the first rocket engine?

In 1938, American James Hart Wyld designed, built, and tested the first regeneratively cooled liquid rocket engine in the United States.

What was the role of aerospace engineering in the Space Race of the 1960s?

Major growth in aerospace engineering occurred during the 1950s and 1960s due to the Space Race. Aerospace engineers were tasked with designing a spacecraft (both manned and unmanned) with the aerodynamics, materials, operating systems, and technology to travel outside of Earth's atmosphere, orbit Earth, explore space, and return to Earth safely.

What is NASA?

The National Aeronautics and Space Administration (NASA) was created in 1958 by the federal government as a nonmilitary entity to lead the

development of aerospace research. NASA is the government agency responsible for science, technology, research, and exploration related to air and space. The highest-ranked official of NASA is given the title administrator, and this person is nominated by the president of the United States. The NASA administrator serves as the senior space science advisor to the president. NASA has research centers and laboratories located across the nation. Interestingly, NASA is known for its astronauts, but they only make up a small fraction of the NASA workforce.

This agency is responsible for more than just space research. Most people are aware of NASA's role in sending astronauts to space and designing space shuttles, but few are aware of their other work. Here are a few other things that NASA does:

- NASA sends satellites to space to orbit our planet, Earth, and learn more about it. This has helped us understand weather patterns.
- NASA sends space probes, robotic spacecraft, to study planets and systems beyond our own. So far, they have visited every planet in our solar system.
- NASA has a program that has a goal of sending humans to explore the moon and Mars. Currently, rovers are best suited for exploration, and humans live and work on the International Space Station. NASA shares their discoveries so that companies can create new spinoff products.
- NASA shares their discoveries and educational materials for teachers to use in classrooms.

NASA is the government organization that handles all science and exploration related to space for the United States (pictured is NASA's Vehicle Assembly Building at the Kennedy Space Center in Florida).

Do all aerospace engineers work for NASA?

DID YOU KNOW?

Most aerospace engineers do not work at NASA. Aerospace engineers work for aircraft and manufacturing companies, many with defense contracts. Many NASA employees are scientists and engineers, but people hold many other positions in this agency.

What do astronauts do in space?

Astronauts have a variety of jobs that they can do, specifically on the International Space Station, which is a permanent laboratory where NASA astronauts and scientists (the crew) live for a predetermined period of time and perform many roles. Six people live on the ISS. They can conduct medical research, develop practices, and test materials that will solve some of Earth's problems and advance technology. The astronauts in the crew may explore space outside of the space station through extravehicular activities, robotics, and experimental conditions and also to make sure the ISS is properly functioning.

What does it take to be an astronaut for NASA?

NASA has some basic requirements to become an astronaut:

1. Astronaut candidates must have a bachelor's degree in engineering, biological science, physical science, computer science, or mathematics.
2. After they earn their degree, they must have at least three years of professional experience or at least 1,000 hours of pilot-in-command time in a jet aircraft. Teachers also qualify as astronaut candidates if they have the correct degree!
3. They must pass the NASA astronaut physical and be able to perform space walks and manage their space suits.

What are other examples of flight vehicles?

A flight vehicle is a type of transportation that is designed to travel in air or space. They can range from simple in design to complex and sophisticated.

They include space shuttles, missiles, satellites, rockets, hypersonic planes, and hot-air balloons. Companies like Uber and Boeing are making a push for future flight vehicles to include air taxis.

What are some future challenges for aerospace engineers?

Aerospace engineers will continue to be needed. Aerospace engineering, however, experiences cycles of activity and inactivity. Humans are determined to explore space. A major challenge is how to get more people in space and, as the frontier expands, how to go faster and farther into space.

What are examples of innovations based on aerospace engineering advances?

- Satellite technology brought expanded television programming, radio relay with satellites, weather forecasting, the connectedness of humanity, Google Maps, cellular service, GPS navigation, the discovery of ozone holes, and the study of climate change.
- Infrared ear thermometers allow temperature to be checked through an ear drum. An infrared ray measures the temperature inside an ear and reduces the potential of infection and cross contamination in hospitals. Infrared radiation is invisible to the human eye, but its energy can be felt as heat. NASA engineers used this technology to measure the temperature of stars.

Your phone map app is made possible by satellites and aerospace engineering. You can thank them the next time you need directions to an unfamiliar house or business.

- The Advanced Research Projects Agency developed the world's first global satellite navigation system, Transit, in 1960. This system determined the location of a satellite from the ground by using radio signals. In 1996, the U.S. Defense Department replaced it with GPS. That technology has been advanced and now is used for autonomous vehicles and on mapping apps used by everyday people.
- Ceramic braces are also known as clear braces or "Invisible Braces." These orthodontic tools are used for teeth straightening and the brackets are made of ceramic. Ceradyne was a company that developed advanced ceramics for defense and aerospace uses. In 1986, 3M Unitek contacted Ceradyne to inquire about a strong transparent material that could be used for orthodontics. This collaboration produced a best-selling orthodontic product.
- Before scratch-resistant lenses were made popular by NASA and before the Food and Drug Administration required shatterproof lenses in 1972, ground and polished glass were the typical materials used to create eyeglasses. Dr. Ted Wydeven of NASA developed a coating that would protect space helmet visors and other aerospace equipment from debris damage. When this technology was applied to glasses and sunglasses, the new scratch-resistant lenses had much longer life then the current plastic lens.

PROTOTYPES AND TESTING

What does the design process look like for aerospace engineers?

The design process is similar across engineering disciplines. A unique aspect of the aerospace engineering process is that it often requires teams of people with many different areas of experience. Additionally, complex flight vehicles are very expensive, and a lot of risk is involved in designing and getting them into the air, so it is not uncommon for some flight vehicle projects to be unsuccessful. Whether or not a project is successful, aerospace engineers use a process to work on aerospace products.

In the first phase, the members of the team complete a market analysis. They must understand customer requirements and specifications.

Once they get this information, it is given to the conceptual design team members, and they make the initial sketches of the design given the customer's needs. As the engineers develop the conceptual design, they can also begin to identify the materials needed for the design and the costs of the materials. They also begin to develop estimates of the conceptual design's performance. Based on the design, the engineers can estimate speed, drag, and takeoff and landing differences. As they design the vehicle and make preliminary decisions, they can modify the design to meet the customer's requirements and specifications.

During phase three, the preliminary design phase, the engineers take the conceptual design, which has been improved iteratively, and begin to develop computer models to predict how well the vehicle will behave under different conditions.

In phase four, the detail design phase, an aerospace engineer takes everything they have learned from conceptual design to virtual modeling and testing and construct a prototype. The team builds each component that is needed for the flight vehicle. They even build full-scale mock-ups from very inexpensive materials such as wood and cardboard. They can also use 3D printers to make parts. In this phase, the engineers also complete physical tests, such as wind tunnel model and aircraft control tests. Because flight vehicles are complex, many different types of engineers can test different components of the design.

Finally, in the fifth and final phase, the prototype is flight tested. In this phase, the engineers work with test pilots to ensure that the flight vehicle operates as designed and is safe. They also ensure that the vehicle meets the customer's requirements and specifications.

This process is not linear, and if at any point they need to go back to a previous design phase, the engineers typically do.

What is a mock-up?

A mock-up is a visual representation of a design that is typically not fully functional. The purpose of a mock-up is for teaching, demonstration, or design evaluation purposes. It is one way to capture the essence of a design idea. Some engineers consider this the cheapest but most valuable visualization of a design. Engineers use mock-ups to quickly get feedback on concepts from a user. It is a very rough version of a prototype, and engineers do not spend a lot of time on it to make it look beautiful and refined. The mock-up should include the bare minimum of resources needed in order to communicate the conceptual design.

What types of materials do aerospace engineers use in their designs?

Aerospace engineers have used 3D printing to build many aspects of aircraft, including air ducts, armrests, seat end caps, seat frameworks, and wall panels. Aerospace engineers also use injection molding; that is, they form specific parts out of hardened plastics (molten, malleable plastic). These hardened plastics become molds that can then be used to manufacture parts that meet the design criteria. Some aerospace parts created with injection molding include turbine housings, turbine blades, and pin maps, which are used to mold carbon or glass fibers into sheet applications.

Do aerospace engineers build full-scale prototypes of their designs?

Aerospace engineers build full-scale prototypes to test and troubleshoot performance and issues with the design, systems, and materials. After the designed part is 3D printed, the testing continues. Engineers perform benchtop tests on the parts in order to find out what makes the part fail and when it will fail. If the part fails in an unexpected way, they go back to the 3D design and rework or modify its features and then reprint and test the part. All the changes are saved so that a record exists of what was done to the part. This is also helpful if the engineer has to go back to a previous design. The engineers are also able to use CAD and 3D printing to scale their designs and print them to the exact specifications and sizes of the parts that will be printed by the manufacturers.

What is a flight simulator?

A flight simulator is a machine designed to replicate the cockpit; it uses computer applications to re-create the instrumentation and conditions of flying. It is used to train aircraft pilots.

What is SPICE? Where was it developed?

SPICE (Simulation Program with Integrated Circuit Emphasis) is an open-source analog simulator that is extensively used by engineers to anticipate the behavior of electrical circuits and, in particular, the integrity of circuit designs. This was created at the University of California at Berkeley.

What is a wind tunnel test?

Aerospace engineers use wind tunnel experiments to simulate the environment around an aircraft in flight. They can study how the aircraft would behave under various conditions and observe its stability and controllability. Using

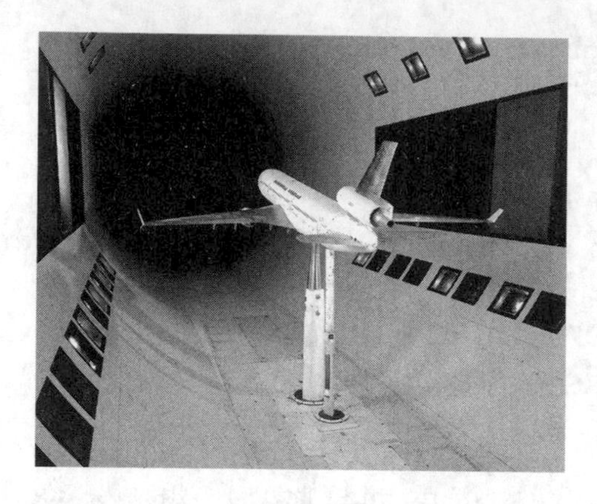

A model airplane is tested in a NASA wind tunnel. Wind tunnels are an aid to testing how vehicles will perform in certain environments.

this machine, the engineers can make controlled testing environments and take measurements of the forces acting on the model aircraft to determine the magnitude of the lift and drag forces on the actual aircraft. The first wind tunnel was built in the 1800s, 30 years before the Wright brothers' Kitty Hawk aircraft.

Who designed the first wind tunnel?

In 1871, Francis H. Whenham (1824–1908) grew unhappy with the whirling arm, which was the mechanical testing approach used at that time. His wind tunnel design used a fan blower, a steam engine, and a tube to push air through. He mounted various parts and shapes and studied their lift and drag. This helped the aerospace industry at the time understand the factors that controlled lift and drag on model aircraft. They later learned that the airflow patterns over scaled models in testing facilities were the same as the full-scale prototypes.

Are all wind tunnels the same?

Each wind tunnel is designed for a specific purpose and to test specific conditions depending on the model to be tested. The speed in the test section depends on the model to be tested and the design of the tunnel.

How did aeronautical engineers perform tests and experiments in the early years?

They initially used nature for testing. They soon realized that they needed a consistent and controlled environment. In the early days, engineers and de-

DID YOU KNOW?

What does transonic mean?

Transonic describes aircraft that have a Mach number M = 1, with speeds that are approaching the speed of sound.

signers used two options: they either moved the aircraft through the air at a specific velocity or moved air past a stationary model of the aircraft. Natural testing sites include windswept ridges and the openings of caves.

Speed!

What is a Mach number?

A Mach number is a dimensionless quantity in fluid dynamics. It is the ratio of an object's speed to the speed of sound. The Mach number is extensively studied in order to have a thorough understanding of the motion of aircraft and rockets. It was named in honor of Ernst Mach, a nineteenth-century physicist who studied gas dynamics.

What does subsonic mean?

Subsonic describes aircraft that have speeds much less than the speed of sound. That is typically less than 250 miles per hour and is designated as Mach number M < 1. One of the first powered, subsonic aircraft was the 1903 Wright brothers' *Flyer I.* Today, most commercial airlines have subsonic aircraft.

What does supersonic mean?

Supersonic describes conditions that are greater than 1 and less than 3. High supersonic describes conditions where the Mach number is greater than 3 and less than 5.

DID YOU KNOW?

When a pilot is assigned a speed, how much can they deviate from that speed?

A pilot can run and indicate plus or minus 10 knots or 0.02 Mach number of prescribed speed by complying with and maintaining the actual speed-adjustment regulations.

What does hypersonic mean?

Hypersonic is five times the speed of sound. Consider that when a spacecraft is reentering Earth's atmosphere, its Mach number is around 25.

What are some famous aerospace equations?

Newton's laws of motion, Bernoulli's principle, and Euler's laws of motion.

How are aerospace equations used in everyday life?

Aerodynamics equations are used to predict the weather in aerospace applications.

CAREERS

How do aerospace engineers use mathematics and science?

An in-depth knowledge of math, physics, chemistry, and materials science is needed in order to resolve common issues with flight trajectory, propulsion, aerodynamic forces, and aircraft design, to name a few. Engineers apply mathematics and science principles to design new aircraft and solve problems that arise in their work.

Where do aerospace engineers work?

Aerospace engineers work for private companies and companies with government contracts designing aircraft, spacecraft, satellites, and missiles. They design and build parts. Some aerospace engineers work at NASA or work with hardware and operating systems. As they advance at NASA, they move to roles in administration.

What are flight-based aerospace products and companies?

Common aerospace products that the general public is very familiar with are flight vehicles. Each of these flight vehicles requires up to 1 million individual parts. Aerospace engineers also provide the services needed to support flight vehicles and maintain them. A specific segment of aerospace products that is also a huge moneymaker is military aircraft. It is very expensive to manufacture aerospace products. Additionally, these products must have the most advanced technology. Therefore, not many manufacturers of aerospace products exist. The aerospace industry is small but mighty and includes companies such as Boeing, Lockheed Martin, and Raytheon in the United States.

Some products that depend on innovations from aerospace engineers include commercial and military airplanes and helicopters; remotely piloted aircraft and rotorcrafts; spacecraft, including launch vehicles and satellites; and military missiles and rockets.

What are some challenges in the aerospace industry?

Challenges faced by the aerospace industry include:

1. Managing aerospace's supply chain:
    - Some vendors find it difficult to meet demands. Suppliers are under immense pressure to assure timely delivery while keeping prices under control as OEMs scale up production to meet their enormous backlogs.

- The supply chains of aerospace companies may be difficult to manage since they rely on hundreds of vendors and subcontractors to purchase raw materials and construct complex products.
- Difficulties occur in achieving the optimum degree of continuous improvement to enable production growth while maintaining quality and keeping costs under control.
- Developing innovations on the shop floor, creating a better consumer experience, deeper engagement, and creating a supply chain system with the precision to fulfill orders.

2. Managing and retaining a diverse workforce:
   - Increasing the workforce with the necessary skills to compete internationally, particularly in the disciplines of science, technology, engineering, and math (STEM).
   - Diverse teams improve an organization's productivity. Also, cross-cultural skills are a common occurrence in the aerospace sector.
   - A lack of diversity and inadequate succession planning have become concerns.
3. Gaining a competitive edge in the aerospace industry
   - International Organization for Standardization (ISO) certifications, International Traffic in Arms Regulations (ITARs), and First Article Inspections (FAIs) are all fundamental criteria for becoming a defense supplier.
   - Other practices include registrations, sustainability efforts, and green-business practices.

Do all aerospace engineers become pilots or astronauts?

Not all aerospace engineers become pilots or astronauts.

Which topics are covered under aerospace engineering courses?

The topics include aerodynamics, flight mechanics, aerospace propulsion, incompressible fluid mechanics, aerospace structural mechanics, introduction to aerospace engineering, aircraft design, space flight mechanics, aircraft propulsion, and thermodynamics and propulsion.

DID YOU KNOW?

What are some applications of aerospace engineering?

Aerospace engineering applications include national defense (security), effective and reliable air travel, enhancement of communication and knowledge dissemination, and growth in consumerism and supply chain globalization.

What are the job responsibilities of an aerospace engineer?

- Exploring and designing new technologies for aviation, spacecraft, and military systems
- Examining any damage or malfunctions in equipment to determine the cause and provide a solution
- Designing, assembling, and testing aircraft and aerospace products
- Establishing quality acceptance criteria for design techniques, quality standards, after-delivery support, and completion dates
- Evaluating the cost and suitability of new projects that are being proposed
- Examining whether the project or product adheres to engineering principles, safety standards, customer needs, and environmental constraints
- Collaborating with the design team to manufacture aircraft and their individual components

Which industries employed the most aerospace engineers in 2019?

In 2019, aerospace engineers held about 66,400 jobs (as per the U.S. Bureau of Labor Statistics). The distribution was as follows:

- Aerospace product and product part manufacturing: 36 percent
- Federal government, excluding postal service: 16 percent
- Engineering services: 15 percent

- Navigational, measuring, electromedical, and control instrument manufacturing: 10 percent
- Research and development in the physical, engineering, and life sciences: 8 percent

Chemical Engineering

What is chemical engineering?

Chemical engineering is a branch of engineering that deals with chemicals, chemical production, and manufacturing materials and products that require chemical processes. This involves developing tools, methods, and techniques for refining raw materials as well as combining, compounding, and processing chemicals in order to create useful products. It applies chemistry, physics, mathematics, biology, and economics concepts to the efficient production, development, transportation, and transformation of energy and materials.

What do chemical engineers do?

Chemical engineers understand many different chemical processes that occur in a laboratory. They use that knowledge to develop practical applications for how products are developed. Once the process is developed, they are also responsible for maintaining those processes. Chemical engineers design products and processes, develop plans for facility layouts, create new tools to improve products and processes, manage processes and facilities, and do research that inspires new innovations. They have many areas of expertise that cover a broad variety of areas.

DID YOU KNOW?

Do chemical engineers still perform experiments?

Yes, chemical engineers must still understand the basic principles of chemistry. Before they design a large chemical engineering plant or process, they will perform a lab experiment of the chemical process. Then, when they understand the process on a micro level, they will better understand how to select the approach innovations, designs, and technologies needed for full-scale production. This helps them identify potential challenges to the process on a small scale and will allow them to make design considerations for those challenges when they scale up.

What are the different branches of chemical engineering?

Chemical engineering is a field that is constantly evolving. People who graduate with chemical engineering degrees may find that long after graduation, they are working in professional roles that may not have existed when they graduated. Chemical engineers are engaging in lifelong learning as they learn and practice new applications for chemical engineering theory. Here are some branches of chemical engineering that an engineer might work in:

- Biochemical
- Biological
- Molecular
- Petroleum
- Process
- Reaction
- Thermodynamics
- Transport phenomena

Is experimental equipment the same as the equipment used in a processing plant?

Experimental laboratory setups and equipment are typically made of different materials than what will be used in a chemical process plant. Before a chemical process or innovation moves to a large plant, a chemical engineer will design a pilot plant to test out the equipment, process, materials, and knowledge that they learned from the laboratory experiment. This will help them successfully move from producing a small amount of a product (ounces) to a large amount of a product (tons).

What types of knowledge does a chemical engineer use to apply laboratory experiments and lessons learned to larger production processes?

In addition to their knowledge of chemistry, a chemical engineer will use mechanical engineering, environmental engineering, physics, and economics to help them design pilot and large-scale chemical production facilities and processes.

Do chemical engineers use the engineering design process?

Yes, they do! They design facilities, processes, and products. This requires them to use their fundamental chemistry knowledge, along with engineering principles, problem solving, and design processes, to meet the desired goal. They have to think like engineers in order to address the challenges that come from translating laboratory experiments and findings to large-scale production facilities.

Who are some famous chemical engineers?

- Arthur Fry (1931–) and Spencer Silver (1941–): Creators of the Post-It® Note, produced by 3M
- Alice Gast (1958–): Sixteenth president of Imperial College, London
- Paula Therese Hammond (1963–): Professor and department head, MIT
- Mae C. Jemison (1956–): American astronaut, doctor, and the first African American in space
- Tomio Wada: Japanese innovator who created the liquid-crystal display (LCD) pocket calculator, produced by Sharp
- Jack Welch (1935–2020): Former CEO of General Electric

Arthur Fry, co-creator of the Post-It® Note, is a chemical engineer.

What is the difference between a chemist and a chemical engineer?

Chemists provide chemistry expertise. Chemical engineers have expertise in the chemical aspects of process operations. Over time, they developed the expertise to develop advanced methods for chemical processing.

What is the relationship between chemical engineering and chemistry?

The study of chemistry as a science began in the 1600s when chemists such as Robert Boyle worked to formulate Boyle's law, whereas chemical engineering became a distinct discipline in the late 1800s when George E. Davis (1850–1907) coined that term. Chemistry is a branch of science that involves the study of substances, their properties, and how they react. Chemical engineers apply scientific knowledge to solve specific problems and tasks. Rather than generating new chemical substances, chemical engineers are designing and constructing plants that perform large-scale chemical processes. Since 1901, 166 laureates have received the Nobel Prize in Chemistry, but only six of them were chemical engineers: Koichi Tanaka, Jon B. Fenn, Kurt Wuthrich, Linus Carl Pauling, William Francis Giauque, and Robert H. Grubbs.

Tell me more about George E. Davis.

George E. Davis (1850–1907) is considered the Father of Chemical Engineering. Davis was an Englishman who authored *A Handbook of Chemical Engineering* (1901). This was a compilation of 12 lectures and over 20 years of work and observations in various chemical industries, and this book was very well received. In this book, Davis defined chemical engineering and the role of the chemical engineers and argued that chemical engineering should be recognized as its own discipline, distinct from chemistry and mechanical engineering. For many years, he worked as an independent inspector and consultant who visited chemical industrial factories across England. As a result of many visits, he recorded observations of common practices, layouts, procedures, and operations. He designed and proposed a chemical engineering course based on his observations. Unfortunately, at that time, these topics were not well received in Britain. The American chemical industries were interested in his work.

According to George E. Davis, what are the fundamentals of chemical engineering?

The fundamentals of chemical engineering include a knowledge of construction materials; steam production and distribution; transporting solids, liquids, and gases; heat transfer; separation processes; evaporation; and crystallization.

Englishman George E. Davis was the first to coin the term "chemical engineering." He was the author of the seminal *The Handbook of Chemical Engineering* (1901).

DID YOU KNOW?

What is a fun fact about George E. Davis?

Because of his work inspecting chemical plants, he grew to strongly dislike waste and inefficiencies. Davis was invested in creating sustainable procedures that avoided developing wasted products and materials.

Who else contributed to early chemical engineering?

In 1908, a group of chemists and engineers founded the American Institute of Chemical Engineers. Although this organization was formed in 1908, the roots of chemical engineering go back to at least a 100-year history of scientific progress, research, and advances in German universities. Chemists like Justus von Liebig (1803–1873) of Germany contributed seminal research and mentored other famous scientists in their labs. The need for chemical engineering rose with the need to understand and advance industrial chemical processes. Research labs like that of Von Liebig nurtured innovative and creative chemical production methods. The United States, like other countries, wanted to be on the forefront of technological advancement and identified chemical engineering as an avenue to advance. Scientists from European laboratories would travel to the United States to establish new research labs there.

Who is Lewis Norton?

Lewis Norton (1855–1893) was a professor of chemistry at Massachusetts Institute of Technology (MIT). He studied the chemical process industry in Germany and recorded many notes. In 1888, Norton designed and offered a chemical engineering course at MIT. The course materials were mostly derived from his notes on industrial practices that he wrote while in Germany.

Who is Frank H. Thorpe?

Frank H. Thorpe (1864–1932) graduated from both MIT and the University of Heidelberg in Germany. When Lewis Norton died, Thorpe inherited the MIT chemical engineering course that Norton developed. Thorpe went on to publish one of the first chemical engineering textbooks, *Outlines of Industrial Chemistry*.

CHEMICAL PRODUCTS AND INNOVATIONS

What are some chemical products that engineers help develop processes for?

Chemical engineers develop production processes for chemicals, fuels, plastics, fertilizer, computer chips, recycling, cleaning water, and pharmaceutical materials.

What types of problems do chemical engineers solve?

Chemical engineers have expertise in both chemistry and manufacturing processes. Chemical engineers work to design and develop chemical processes

One of the areas in which chemical engineers work is designing and testing chemical plants, as well as making sure that built plants maintain high efficiency.

and plants. They ensure that the chemical processing plants can produce large amounts of chemicals efficiently. This work includes design pilot studies to test plant and process designs, collecting and analyzing data, applying fundamental chemistry knowledge to applied problems, and making informed educated guesses about the outcomes of their designs. One major challenge that chemical engineers solve is how to manage the different parts (unit operations) of a chemical processing plant and their diverse operating conditions. Another problem that chemical engineers solve is called process engineering. Like a chemical plant that has unit operations, chemical reactions can also be segmented into unit processes. The study of unit processes of chemical reactions is called process engineering.

What are the top 10 contributions that chemical engineers have made to society?

- Fueling the world's economies
- Creating cleaner energy
- Products for growing populations
- Removing harmful sulfur from fuels
- Better living through chemistry
- Stretching natural resources
- Large-scale productions engineering
- Convenient and abundant food
- Healing diseases and extending life
- Powering the personal computer

What are some other applications of chemical engineers beyond chemical plants, chemical processing, and oil?

Chemical engineers can work in many different industries. They typically work with people from very diverse disciplines, even outside of engineering. The following industries represent some places you might expect to meet a chemical engineer:

- Computer chip manufacturing and design
- Food industries

DID YOU KNOW?

What scientific advances led to the profession of chemical engineering?

During the Industrial Revolution, engineers were looking for larger-scale applications and utilities of chemicals and materials. This led to the industrial application of chemistry such as the work of scientists like Justus von Liebig and those who came from his laboratory. Also, the study of separation science allowed for scientists to investigate how to separate components from mixtures (study reactions). This was useful in the oil refinery and chemical industries. These forms of chemical processes were used to produce glass and soap from soda ash in Britain in the early 1800s. For industrial applications, an increased understanding and exploration of the processes within which chemical reactions take place needed to occur. This exploration required chemistry, physics, and mathematics.

- Environmental companies
- Government agencies
- Law firms
- Banking

Where do chemical engineers work?

Chemical engineers work in many different types of industries. They might be in a business office, research laboratory, pilot plant, or industrial process facility working in pharmaceuticals, construction, petrochemicals, food processing, electronics, plastic or polymer manufacturing, water treatment, or environmental health and safety.

Who discovered the process of investigating chemicals and their properties?

Chemical engineering emerged from a need to have large-scale facilities that could produce massive quantities of materials and chemicals. Once those facilities were designed, chemical engineering was also needed to design procedures that would ensure the safe design of manufacturing plants. Because of the Industrial Revolution, professional inspectors were required to evaluate the conditions of industrial chemical plants. After many years of inspecting chemical plants, the inspectors were able to create a much needed curriculum to help future chemical engineers understand the conditions of these plants and how to improve the design of future chemical plants.

What are some everyday products and needs that require chemical engineering?

Chemical engineering involves the processing of materials and products to meet specific needs. It is important to highlight how some specific needs are addressed by applying chemical engineering principles. Here are six example problems and six chemical engineering solutions:

Engineering Problem	Chemical Engineering Solutions
Long-lasting lipstick	12-hour lipstick at every drugstore
A new AIDS drug	Enough AIDS drugs for an epidemic
Cleaner rivers	Zero-effluent pulp mills
Faster bone repair	"Instant Grow Protein" for bones
Delicious, at-home pizza	Pizza factory with a special freezing process
Cleaner air	Scrubbers that remove chemicals from factory outputs

What are some examples of innovation in chemical engineering?

Chemical engineers create and contribute innovative solutions that address food scarcity, water transportation, and providing consistent energy resources to more communities. The innermost circle represents some of the different roles and job functions that a chemical engineer might have. The second circle represents some of the opportunities that a chemical engineer has to create innovative chemical engineering solutions in their job. The outermost circle provides specific examples of how a chemical engineer could innovate in their role at work. For example, a chemical engineer who is working in an engineering research and development (R&D) role might innovate by creating a new production process. More specifically, they could create a new piece of technology to improve the process, produce a new product, or improve an old product.

Smokestack scrubbers remove dangerous pollutants in industry chimneys before they contaminate the atmosphere. This is just one of the many creations of chemical engineers that improve our lives.

What is materials processing?

Materials processing takes existing materials and modifies them to have desired chemistry characteristics and properties. A common material that is processed is copper. Materials processing allows chemists to take materials and modify them to create coins. The U.S. Mint has been responsible for producing coins for over 200 years. Let us explore how the U.S. Mint uses materials processing to produce coins. When the U.S. government first began striking coins in 1793, the penny was made of 100 percent copper! Copper costs have increased throughout U.S. history, so it became expensive in terms of both money and energy to process copper. During World War II, copper was an essential material for war efforts. Chemists used materials processing to explore different materials compositions. Since 1982, a penny consists of 97.5 percent zinc and 2.5 percent copper.

CHEMICAL ENGINEERS AND CHEMISTRY

How do chemical engineers use chemistry?

Chemical engineers use the fundamentals of both chemistry and engineering to develop and improve processes that involve materials. They use

chemistry during chemical operations and to develop solutions to technical problems—for example, how to design an insulin-injection pump that more accurately delivers insulin or discovering materials for achieving fast-speed internet. Fiber-optic cables are run throughout the United States, and this allows for many types of data to be transmitted very rapidly. But how can we accomplish such a feat? Chemical engineers had to develop a process to fabricate the fiber-optic cables. Before this new process, the properties of the thin, glass fibers of the cable were very brittle and fractured easily. Chemical engineers developed a specialized coating to prevent the negative properties and maintain the properties that helped transmit data.

What is fundamental research?

Fundamental research is sometimes called basic research. Often, this means that scientists want to expand what is already known about a specific topic. Consider fundamental research like using building blocks to build a structure. First you build the base of your structure and then expand upward or outward as you add more blocks.

What are chemicals?

Atoms, the smallest form of matter, cluster together in different combinations to form compounds. Each atom has a specific charge as identified by the number of protons (positive charge), neutrons, and electrons (negative charge). Atoms of the same type form elements. Chemicals are formed when a reaction occurs that causes atoms to change.

What is the difference between fractionation and distillation?

Both fractionation and distillation are procedures for separating the components of a solution depending on their melting points. When the boiling points of chemicals in a combination vary, distillation is used; when the boiling points of chemicals in a mixture are close to each other, fractionation is utilized.

Other differences are as follows:

- In the complete process of simple distillation, only one distillation (vaporization–condensation) cycle will occur, but the fractional cycle will include at least two cycles.

- Fractional distillation requires an extra apparatus known as a "fractionating column," whereas basic distillation does not require any such apparatus.

What is the periodic table?

A periodic table is used to organize chemicals according to their identified properties. A chemical can be identified by how it behaves and its properties. Some chemicals may behave similarly to other chemicals or have similar properties (or behaviors) as other chemicals. For a long time, scientists have been exploring and observing the chemical properties of elements. Scientists developed a tool to help organize chemicals according to their properties. The periodic table of elements is a table that represents how similar elements are grouped.

What is an example of a chemical that chemical engineers work with?

The chemicals that chemical engineers might use depend on the problem they are trying to solve. Chemical engineers today are most commonly considered process engineers. A process engineer is responsible for the design and operations of processes that produce products. They typically work with chemicals, industrial gases, rubber, soap, fiber, glass, metal, and paper.

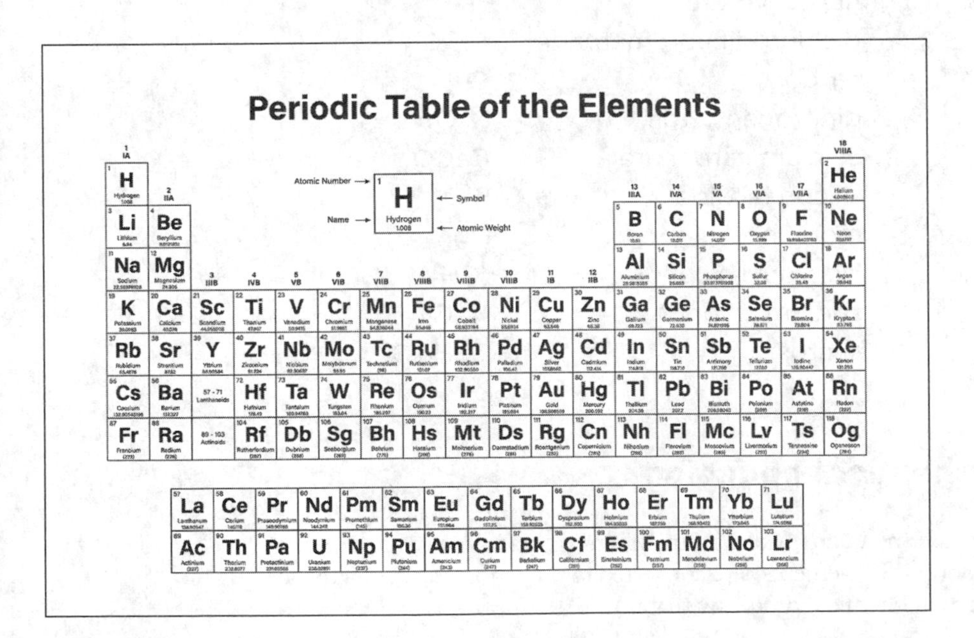

The periodic table of elements organizes all the known elements according to their atomic numbers.

How do chemical engineers reduce the impact of harmful chemicals on the environment?

They will often use catalysts to speed up chemical reactions. Catalysts are chemical compounds, substances composed of two or more elements that have been chemically combined through chemical reactions that speed up chemical reactions and lead to specific products made from a specific chemical reaction. Some catalysts that we use in everyday life include:

Common Name	Chemical Name	Formula
Baking Soda	sodium hydrogen carbonate or sodium bicarbonate	$NaHCO_3$
Bleach (liquid)	sodium hypochlorite or hydrogen peroxide	$NaClO$ or H_2O_2
Borax	sodium tetraborate decahydrate	$Na_2B_4O_7 \cdot 10H_2$
Brimstone	sulfur	S
Cream of Tartar	potassium hydrogen tartrate	$KHC_4H_4O_6$
Epsom Salt	magnesium sulfate heptahydrate	$MgSO_4 \cdot 7H_2O$
Freon	dichlorodifluoromethane	CF_2Cl_2
Graphite	carbon	C
Laughing Gas	dinitrogen oxide	N_2O
Marble	calcium carbonate	$CaCO_3$
Quartz	silicon dioxide	SiO_2
Quicksilver	mercury	Hg
Rubbing Alcohol	isopropyl alcohol	$(CH_3)2CHOH$
Salt	sodium chloride	$NaCl$
Sugar	sucrose	$C_{12}H_{22}O_{11}$

What is a chemical reaction?

Chemical reactions occur when atoms change or are rearranged. Substances (i.e., compounds or elements) are converted to new substances. Chemical reactions are a form of chemical processes that take place throughout our every-

day life. For example, did you know that the process used to determine how to run the laundry washing machine was determined by understanding a chemical process? Think about this: someone had to determine how much and what type of laundry detergent would be needed to remove dirt from a football player's jersey after a game in the mud and rain! Most chemical reactions are often very difficult to reverse because during the reaction, new substances are formed.

What is a mixture?

A chemical mixture is the combination of elements to form a substance that does not require a chemical reaction. It can be easily separated back into its original components. Mixtures can be undone or reversed. Examples include salt water, trail mix, sand water, salt, and pepper.

What is another example of a mixture?

Hydrocarbons are organic compounds (mixtures) composed of two elements: hydrogen and carbon. These hydrocarbons can be formed in nature, mostly being found in crude oil. Decomposed matter over a long period of timc (thousands and thousands of years) allows for an abundance of hydrogen and carbon to be formed. The simplest hydrocarbon is methane, CH4. Hydrocarbons are often used for fuels like gasoline and wax like candles.

Where do chemical reactions take place?

Chemical reactions take place all around us, but the chemical reactions that chemical engineers use first take place in a laboratory and then are scaled up and mass produced in a manufacturing facility. In this facility, a chemical reactor is used to make a chemical reaction work.

Which chemicals are responsible for the unpleasant odors emanating from wastewater treatment plants?

The unpleasant odors emanating from wastewater treatment plants are caused by the following compounds:

- Hydrogen sulfide
- Methyl mercaptan
- Dimethyl sulfide
- Dimethyl disulfide

Do some common chemical reactions exist right out in the open? What are some examples of chemical reactions?

- Photosynthesis: A process that plants use to produce food. Also, this process converts oxygen to carbon dioxide. That plant in your house uses photosynthesis.
- Combustion: Have you ever lit a match? Combustion is the process of combining hydrocarbons with oxygen. Whenever you use propane in the grill or fireplace, burn a candle, or build a fire, you are right in the middle of a chemical reaction.
- Eating food: Did you know that when you eat food, you are initiating a chemical reaction (participating in a chemical reaction)? One such reaction is the role of saliva in breaking down your food into simpler substances that your body can manage.

What are some examples of activities that chemical engineers will do during lab experiments?

- Mixing
- Heating
- Contaminant control

What are some different ways to measure chemicals?

Chemical engineers use a variety of tools that are dependent on their job function and the specific task they are trying to complete. Some of the tools they use when working with chemicals include:

- Analyzers: to obtain samples from chemical processes
- Chromatography columns: to separate mixtures
- Colorimeters: to measure light changes in a sample
- Conductivity meters: to measure how well a solution can carry an electric charge
- Detectors and monitors: to determine if hazardous gas is present during a process
- Humidity measures: to determine how much water vapor is in the air

PROCESSING PLANTS

What is a pilot plant?

A pilot plant is a facility that allows a chemical engineer and their team to test new technologies before a product is produced commercially. It allows for small batches of the new products to be produced so that the chemical engineer can learn about the new technology. It essentially allows engineers to experiment with different aspects of chemical production processes on a realistic but small scale.

What is a chemical plant/factory?

A chemical plant is a facility that is designed to house a desired chemical process. New materials are created by performing various unit operations. To perform these operations, the plants are filled with special equipment and technology and must follow very high safety standards. A chemical plant is a type of industrial process plant that blends (manufactures) chemicals on a large scale. They are typically located near the source of the raw materials that will be processed. Typical products from chemical plants include:

- Pharmaceutical
- Agrochemical
- Food and beverages
- Sheet material
- Textiles
- Treated gases

DID YOU KNOW?

What is the difference between process design and plant design?

Chemical engineering involves changing the physical properties of materials. This process often involves using materials in bulk, which means that a facility must exist to house such large quantities. Chemical engineers must design processing plants that will carry out the processes to do large-scale materials processing. The two key activities are designing the actual process to change the materials and designing the plant layout that will accomplish the processing.

How is a chemical plant designed?

A chemical processing plant is designed in stages. First, the chemical engineer will determine the different approaches/processes they could use to create the desired product and do a feasibility analysis of each process. Often, chemical engineers know how to make the product in a laboratory, but that process must be scaled up so that the product can be made available to many more people. Selecting the best process requires the chemical engineer to design the process, test the process, collect data on how well it worked, and calculate how much it might cost to replicate.

What is the anatomy of a chemical engineering processing plant?

A processing plant has many parts, which include everything from storing substances and materials to safety elements such as ventilation systems.

How does a chemical engineer determine the layout of a plant?

Once a chemical engineer has identified the process required to complete a chemical process, they will use any of the following approaches to make decisions about the plant layout. They may use a combination of these approaches.

Every approach they use must be one they have confidence in and can be replicated and reviewed by other chemical engineers.

- Intuition based on experience: uses experiences and knowledge based on prior successful designs
- Economic optimization: considers ways to reduce the distance that materials have to travel within the plant
- Rating: develops a way to score the different components in the plant. This will help explain relationships between equipment, materials, and hazards.
- Mathematical modeling: uses mathematical equations to make decisions about where to place equipment and materials in the layout.
- Software-based approaches: designs the layout in 3D computer-aided design software.

UNIT OPERATIONS

What are unit operations?

Unit operations are the basic steps in a series of tasks that must be completed to meet a goal or output. In chemical engineering, unit operations involve a material enduring a physical or chemical change that requires various tasks such as:

- Pulverizing
- Mixing
- Heating
- Roasting
- Absorbing
- Precipitation
- Crystallizing
- Filtering
- Dissolving

How are unit operations organized?

Each unit operation represents a specific task that is part of a larger process. The extensive list of unit operations can be reduced to five classes. The processes can be a pure-unit operation (that is, it can fit into one of these categories), or it can be a mixed-unit operation (it is a combination of multiple processing categories):

- Mechanical processes
- Mass-transfer processes
- Fluid-flow processes
- Thermodynamic processes
- Heat-transfer processes

What are mechanical processes?

Mechanical processes include a variety of operations that physically change a material involved in an operation. The mechanical processes include crushing and pulverization of materials to small, sand-size pieces, size enlargement, and separating materials with mechanical or physical forces.

What are mass transfer processes?

Mass transfer processes occur when mass is transferred from one point to another, which causes distillation, gas absorption, extraction, adsorption, and drying. The evaporation of water from a pond to the atmosphere, the filtration of blood in the kidneys and liver, and the distillation of alcohol are all examples of mass transfer processeses.

What are thermodynamic processes?

Thermodynamic processes are typically changes in pressure, volume, internal energy, temperature, or any sort of heat transfer. For example, in refrigeration, a warmer object is made cold.

What are heat transfer processes?

Heat transfer processes refer to a difference in heat and the rate it moves from one point to another. Consider some common processes like condensation (gas changing to liquid; think of blowing human breath on a car window, and liquid forms because the car window is cooler) and evaporation (liquid changing to vapor; think of drying wet clothes in the summer sun).

DID YOU KNOW?

What is a fluid flow? What are fluid flow processes?

Fluid flow is a branch of fluid mechanics that deals with fluid dynamics. The motion of a fluid subjected to uneven forces is involved. As long as unbalanced forces are applied, this motion will persist. Fluid flow processes include filtration, fluid transportation, and solid fluidization.

How do chemical engineers use unit operations to build chemical plants?

The parts of a chemical processing plant are called unit operations. They are the building blocks of a chemical plant and represent the physical aspects of chemical engineering. In fact, early chemical engineers believed that unit operations could be used in a variety of industrial processes. In the case of the chemical engineering plant, the unit operations are the coordinated series of steps used in a chemical process.

What are chemical process unit operations?

Process unit operations require a collection of unit operations to achieve a desired goal or develop a specific product. An example of a process unit operation is milk processing. Once the milk arrives at the plant, the unit operations included in this process are homogenization, pasteurization, chilling, and packaging.

PROCESS OPERATIONS

What is process control?

Process controls is a specialty area within chemical engineering that includes physics, chemistry, biology, algorithms, and mechanisms. Process controls ensure that any process that needs continuous monitoring is well controlled. In the past, a process operator and process engineer were responsible to monitor

all chemical processes and operations. Chemical plants now include various levels of automation.

What is a digital control system?

The engineers and operators use digital control systems (DCS) that automate and manage plant processes. Consider that before digital control systems, if the device had a valve that required turning, the process engineer had to manually turn the valve. With the help of DCS, processes can be controlled remotely and the engineers. From the control system, engineers can retrieve data on how well the process is operating and even remote control the operation.

What are the benefits of process control?

Chemical plants have a great potential to cause harm if there is a system failure. Process control helps to maximize safety of workers, reduces costs, and reduces the negative impact of chemical processes on the environment.

What is an example of a failure in process control?

On December 3, 1984, the Bhopal Gas Tragedy killed thousands of people and injured an estimated 500,000 people. About 40 tons of gas leaked

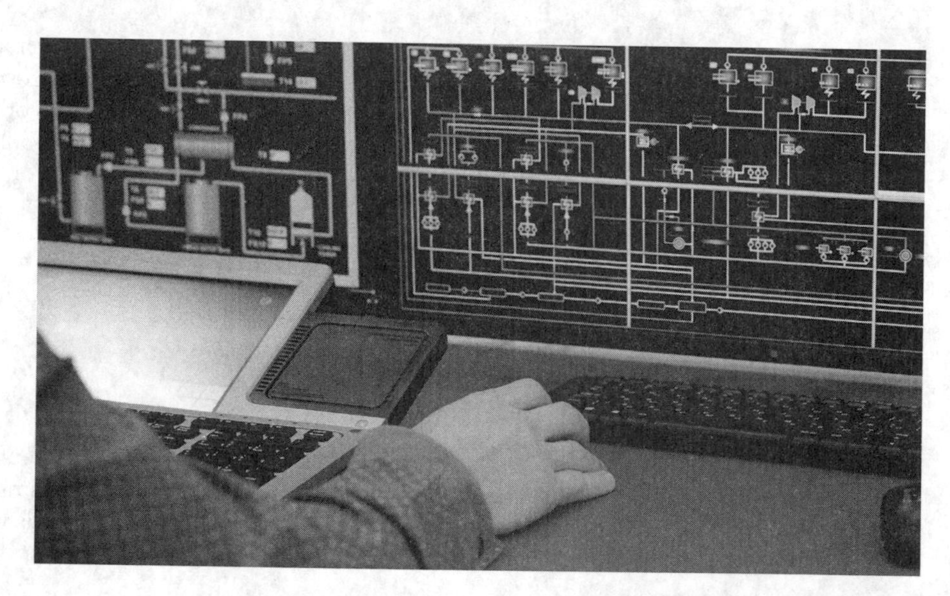

A digital control system helps operators to manage all of a plant's processes through one electronic interface.

from a pesticide plant in Bhopal, India. A runaway chemical reaction occurred in the mixing tank that caused temperatures and pressure to rise so high that it forced emergency venting of toxic gases into the surrounding community. There were unfortunately several controls that were not operational at the time of failure and could have prevented the tragedy.

CHEMICAL REACTION ENGINEERING

What is chemical reaction engineering?

In chemical reaction engineering, the engineer is responsible for taking chemical reactions from laboratory scale to commercial (large-production) scale. They also design and operate chemical reactors and improve industrial-scale catalysis and various manufacturing processes. They can also introduce new technologies needed to generate chemical reactions. Reaction engineering can have an impact on environmental pollution and drug delivery within the human body.

What is a chemical reactor?

A chemical reactor is a vessel used to house a chemical reaction. The most common modes of operations are batch, continuous stirred-tank, plug flow, and semibatch reactors. In a chemical plant, the size of the reactors and the mode of operation determine how much the plant is able to produce. Batch processes are best applied for small production rates and flexibility, while continuous process modes allow for large daily production.

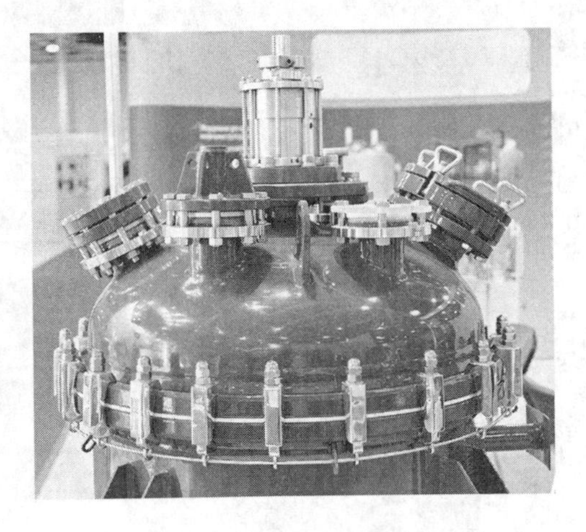

Chemical reactors are designed to contain chemicals and control their interactions safely.

Do different types of chemical reactors exist?

Chemical engineers use different tools to do their work. These are not items you would see in everyday life. As you can imagine, chemical engineers work with many different kinds of chemical reactors. Here are some examples:

- Plug flow reactors
- Batch reactors
- Continuous stirred-tank reactors (CSTRs)
- Tricklebed reactors

THERMODYNAMICS FOR CHEMICAL ENGINEERS

How do chemical engineers use the fundamentals of thermodynamics?

Thermodynamics is the study of the conversion of energy and how that conversion effects or can be measured by changes in temperature, pressure, and volume. Chemical engineers use thermodynamics to understand how heat behaves in a chemical reaction and in chemical reactors. This influences process plans and reactor designs.

What are the laws of thermodynamics, and why are they important?

The laws of thermodynamics are important because they help us understand how materials behave. It also gives chemical engineers insights into which reactions will work and which ones will not work.

- The first law of thermodynamics, also called the law of conservation of energy, states that energy cannot be created or destroyed; it only changes forms and moves around.
- The second law of thermodynamics states that the entropy (measure of disorder) of any isolated system either stays the same or increases;

it never decreases. One example is a broken glass. Once it is in many thousands of pieces, it can never be put back together again. Consider the brokenness as disorder; the disorder increased in the glass when it broke.

- The third law of thermodynamics states that the entropy of a system approaches a constant value as the temperature approaches absolute zero. This means that if you can stop a group of atoms cold enough to freeze in position, they will no longer move and be disorganized. From this law, chemical engineers learn that motion, temperature, and disorder are related to one another.

TRANSPORT PHENOMENA (TRANSPORT PROCESSES)

How do chemical engineers use math to understand how chemicals behave in chemical reactions?

It is important for chemical engineers to have a strong understanding of how heat behaves and how fluid will move or flow in chemical reactions. They use advanced mathematical questions to model physical processes that involve the transport of substances from very small (molecular) levels to very large (macroscopic) levels.

What is an example of a transport process?

One example of molecular transport is something that most people experience at some point in their lives. Imagine playing a sport and rolling your ankle, and suddenly, you realize that it is swollen. At this point, some people will soak their swollen ankle in Epsom salt. As the ankle is soaking in the Epsom salt and water, the water moves through your skin to carry the Epsom salt, which will eventually help to reduce swelling.

SEPARATION PROCESSES

What is a separation process?

A separation process is used when a mixture of substances is converted into two or more distinct products. Many different techniques can be used to separate mixtures at different stages of production processes. Separation processes can occur either in the laboratory or the process plant. One example of a separation process is distillation. One example of a product that must go through a separation process is crude oil.

What is crude oil?

Crude oil products are typically made of gas, oil, water, and contaminants that must be separated and then processed for use in cars and homes and for construction purposes. Crude oil is a hydrocarbon found in deposits deep underground that formed from aged animal and plant matter. With the rise of the automobile industry came the increased and rapid growth of dependence on fuel. In this natural state, crude oil contains many molecules, or components of carbon and hydrogen atoms. Crude oil cannot be used in its natural state to power cars and vehicles. For example, some of its molecules are heavy and can clog a fuel system.

What is oil refining?

Crude oil in its natural state is not ideal for fuel systems. All the diverse molecules must be separated to meet specific needs. Therefore, chemists and

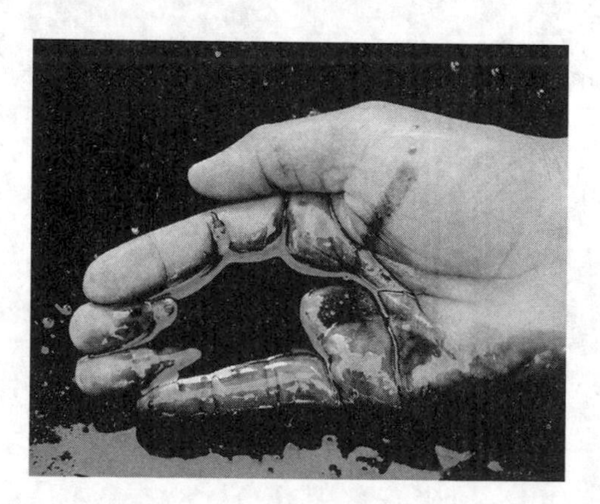

Crude oil is the raw hydrocarbon fluid found in underground deposits. It needs to be refined into other forms such as gasoline before it is useful.

chemical engineers had to find the "just right" composition of molecules that could power a fuel system. For example, for automotive fuel systems, the molecule should consist of 5–10 carbon atoms and 10–25 hydrogen atoms. Different combinations of carbon and hydrogen atoms in crude oil make up the many diverse molecules that are used to provide chemical energy. Chemists explore and research each of these molecules to understand their characteristics, behavior, and application. For example, isooctane molecules, which are ideal for

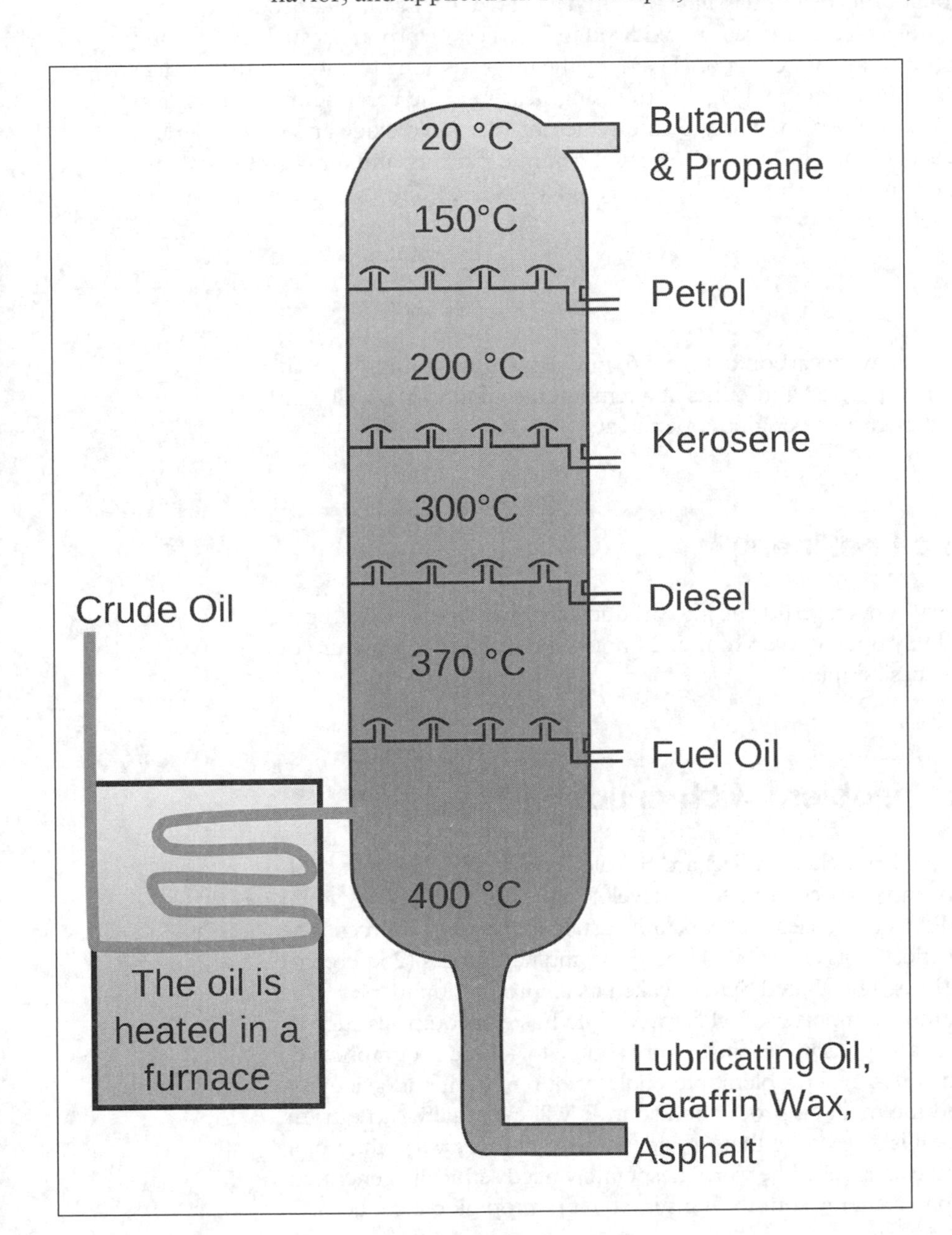

Figure 18: Crude oil is refined using fractional distillation, meaning it can be separated into different products like gasoline, kerosene, and fuel oil because these fuels have different boiling points.

DID YOU KNOW?

What are flameless oxidizers?

Flameless oxidation is a thermal treatment that combines waste gas, ambient air, and auxiliary fuel before passing it through a heated inert ceramic media bed. The organic molecules in the gas are oxidized to harmless byproducts, such as carbon dioxide (CO_2) and water vapor (H_2O), by the transmission of heat from the media to the gaseous mixture while also releasing heat into the ceramic media bed. To start the decomposition of organic molecules into carbon dioxide and water, flameless oxidizers employ electrically heated ceramic packing and a high-velocity introduction mechanism. Furthermore, once this oxidation process starts, it self-perpetuates.

automotive gasoline, have a carbon count of 5 and a hydrogen count of 18. It is ideal because it is vaporized and burns at a temperature that works well with the temperatures that gasoline engines operate at.

What is an oil refinery?

An oil refinery converts crude oil into products that we can use—example, fuel for vehicles. They operate year-round, 24 hours a day. Here is a picture of an oil refinery in Mississippi:

What is the problem with crude oil?

In 1996, Nobel Prize chemist Richard Smalley considered the great challenge facing the twenty-first century to be developing technologies that would support the world's access to clean, cheap, and sustainable energy sources. The United States gets most of its crude oil (41 percent) and natural gas (25 percent) from the state of Texas. The United States peaked its oil production in the 1970s and has now become an importer of oil from Middle Eastern countries such as Saudi Arabia. As oil and gas sources move from source to source geographically, the world is faced with a real problem. No endless source of oil and gas exists on this planet, and the main oil production sources will eventually run out. As you can imagine, oil is a resource that provides power in more ways than one, both politically and electrically. The world desperately needs ethically generated, cheap, and sustainable energy sources. This is why it is a top global concern.

How much energy does the world use?

On average, the world consumes 97 million barrels of oil each day. This value has been increasing every year since 2010. Remember, each barrel holds 42 gallons of oil. One of the products we get from a barrel of oil is gasoline, about 19 gallons per barrel. The majority of U.S. vehicles require gasoline. This means that we use a huge amount of oil, and a high demand exists for energy. Unfortunately, not all people around the world have equitable access to energy, but if every person in the world (about 10 billion) was able to access just 2 kilowatt-hours of energy each day, then about 900 million barrels of oil would be needed each day. This would utilize about 60 terawatts of energy around the entire planet.

What are oxidation reaction, reduction reaction, and redox reaction?

The phrases oxidation and reduction refer to the addition or removal of oxygen from a substance., i.e., when a loss of hydrogen electrons or gain of oxygen occurs, it is known as oxidation reaction, whereas when a gain of hydrogen electrons or loss of oxygen occurs, it is known as a reduction reaction. For example: The iron in metal is oxidized, resulting in the formation of iron oxide, sometimes known as rust. Meanwhile, oxygen levels are reduced. This is an example of an oxidation reaction. On the other hand, iron is formed when iron oxide is reduced. This is an example of a reduction reaction.

An oxidation-reduction (redox) reaction is a type of chemical reaction in which two species exchange electrons. Any chemical process in which the oxidation number of a molecule, atom, or ion changes by gaining or losing an electron is known as an oxidation-reduction reaction. Redox processes are frequent and necessary for some of life's most fundamental operations such as photosynthesis and respiration.

What can be produced from a 42-gallon barrel of crude oil?

Here is what can be produced from a single 42-gallon barrel of crude oil. Although it is called a 42-gallon barrel, it can actually store 45 gallons of substance.

Product	Gallons
Finished motor gasoline	19.53
Distillate fuel oil	12.18

Product	Gallons
Kerosene-type jet fuel	4.16
Petroleum coke	2.23
Still gas	1.68
Hydrocarbon gas liquids	1.55
Residual fuel oil	1.05
Asphalt and road oil	0.80
Naptha for feedstocks	0.46
Lubricants	0.42
Other oils for feedstocks	0.29
Miscellaneous products	0.25
Special napthas	0.08
Finished aviation gasoline	0.04
Kerosene	< 0.01
Waxes	< 0.01
Total	44.73

What is a threshold limit value?

The American Conference of Governmental Industrial Hygienists, Inc. (ACGIH) establishes the threshold limit value, which is defined as the amount of chemical substance concentration at which a person can operate without an unreasonable risk of disease or damage. It can be measured in parts per million (ppm) or milligrams per cubic meter (mg/m3).

What is a wet-bulb globe temperature? Where is it used?

A wet-bulb globe temperature (WBGT) is a form of apparent temperature that is used to assess how temperature, humidity, wind speed, and visible and infrared radiation (typically sunlight) affect humans. Industrial hygienists, athletes, athletic events, and the military utilize it to assess safe limits of exposure to high temperatures. Thus, it is used to measure the sultriness of the environment. It consists of three temperature readings, namely wet-bulb temperature, ordinary dry-bulb temperature, and black-bulb globe temperature.

How do chemical engineers help solve the world's energy problems?

Chemical engineers and other engineers are needed to address sociotechnical problems all over the world. Our world needs a non-oil-based solution that will generate at least 1 terabyte of energy from natural, sustainable sources. We need to fully explore the boundaries of alternative-energy sources. They need to use chemical engineering knowledge to discover the "new oil" of the twenty-first century that would be efficient, safe, and equitable.

What alternative-energy solutions do chemical engineers explore?

Hydrogen is a common solution to alternative-energy needs. It is used for fuel cells, which was at one time one of the most popular solutions that automotive companies considered for alternative-energy vehicles. The challenge with hydrogen is that it must be produced, stored, transported, and charged, whereas with gas, we have figured out all of that after over 100 years of investigation. It can be expensive not only to generate but also to transport. It must be very carefully stored. This is very similar to electricity.

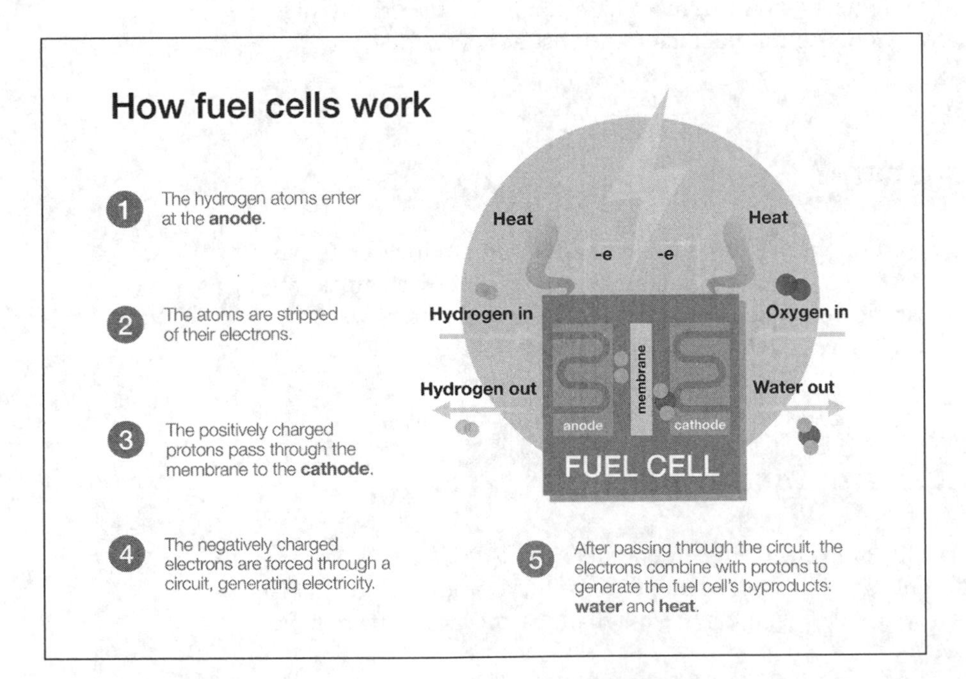

How fuel cells work
1 The hydrogen atoms enter at the anode.
2 The atoms are stripped of their electrons.
3 The positively charged protons pass through the membrane to the cathode.
4 The negatively charged electrons are forced through a circuit, generating electricity.
5 After passing through the circuit, the electrons combine with protons to generate the fuel cell's byproducts: water and heat.
Heat
Heat
-e
-e
Hydrogen in
Oxygen in
Hydrogen out
Water out
membrane
anode
cathode
FUEL CELL

Figure 19: Hydrogen fuel cell technology is being used already in some vehicles such as public buses. It is one alternative form of energy chemical engineers are developing.

DID YOU KNOW?

What does it mean for a system to be stable?

A chemical system needs to be highly elastic, meaning that if a disturbance occurs, it will bounce back and return to good operations; losing control and unpredictability are unwanted when disturbances occur. A stable system means that a bounded output occurs for a bounded input.

PROCESS AND DYNAMICS CONTROL

What is process control?

Process control describes the methods that engineers use to monitor, adjust, and correct when changes occur in processes and systems. Engineers in this role must have a strong understanding of other process operations, unit operations, reaction engineers, and thermodynamics and how to apply high levels of mathematics modeling to understand the changes occurring.

What is a system?

Systems are based on inputs and outputs. System inputs include things that cause changes. System outputs are things that are affected by changes (responses). Outputs can be controlled variables, which include pressure, temperature, flow rate, liquid level, or material composition.

What are system dynamics?

System dynamics are very useful in engineering. Sometimes, systems exhibit slight differences in behaviors that are difficult to understand. Engineers have found ways to understand system dynamics and how they behave in time.

What are some tools that chemical engineers might use in process control?

Processes and systems are always changing as substances move through different unit operations during production. In addition to knowing chemical engineering fundamentals and high levels of mathematics, engineers also use technology to help them control the system that they have developed a mathematical understanding of. This could include the use of sensors to collect real-time information more quickly and efficiently. They also use process control panels that illustrate the entire process and provide access to current levels and activity that the engineer can use to make adjustments to the unit operations.

THE FUTURE OF ENGINEERING AND ITS CHALLENGES

What are some future challenges for chemical engineers?

The challenges that chemical engineers will have to develop solutions for relate to energy, the environment, water, health, and nutrition. The demand for energy is at an all-time high. The average total consumption is 16 terawatts per hour. That is 16 trillion watts of power. Imagine 160 billion 100-watt light bulbs being used every hour! Unfortunately, fossil fuel (the most common source for much of our energy) reserves are significantly reduced. Fossil fuel sources include oil, natural gas, and coal. This supply is not sustainable and nonrenewable, and when it runs out, it will be gone forever. Given all the diverse energy needs and the great energy demand across the globe, not one single solution exists to provide the energy needed. This is a chemical engineering problem of the future.

Water chemical engineering challenges include access to water, ensuring consistent water quality, and water energy and its footprint. Chemical engineering solutions to water quality and purity include UV, adsportion columns, ELISA, irrigation, and wastewater treatment. Chemical engineering healthcare challenges include flu pandemics, increased rates of cancer, AIDS, malaria, increased autoimmune diseases, and rising healthcare costs.

What are some major engineering challenges?

- Upgrading infrastructures to help maintain the functionality of roads, bridges, water, electricity, and sewage systems

- Educating first-world engineers to learn innovative problem-solving techniques
- Adopting green engineering in manufacturing to enhance sustainability and minimize carbon emissions
- Seeking alternative-energy sources that are feasible
- Creating awareness and encouraging students to opt for STEM education (science, technology, engineering, and mathematics), which plays a critical part in the U.S. economy's long-term development and stability
- Keeping our personal information and assets safe from cyberattacks
- Using engineering innovations to protect the environment and reduce carbon emissions

What is the future of chemical engineering?

Over the last 200 years, chemical engineering has advanced, and its impact has been expanded to other industries. The future of chemical engineering is built upon the fundamental skills of chemical engineers. The future application of these fundamentals can be found in the following industries:

- Microelectronics
- Pharmaceutical/health care
- Biotechnology/bioengineering
- Nanotechnology

How does someone become a chemical engineer?

Although chemical engineering has roots in chemistry and mechanical engineering, once someone graduates from high school, they can go to college and earn an undergraduate degree in chemical engineering. An undergraduate bachelor's degree in engineering typically lasts 4–5 years. During their college years, students will take many classes and will also work on internships and other engineering work experiences.

What are the main chemical engineering courses in universities?

Universities across the country and the world agree on the fundamentals of chemical engineering but may have different approaches to teaching the subject. Some of the main chemical engineering courses include:

- Chemical engineering analysis
- Process modeling and computation
- Chemical engineering thermodynamics
- Fluid mechanics
- Heat and mass transfer
- Chemical reaction engineering
- Chemical process dynamics and control
- Separation processes
- Computer-aided design
- Chemical engineering laboratories

What are some examples of chemical engineering organizations?

- Institution of Chemical Engineers (IChemE)
- American Institution of Chemical Engineers (AIChE)

What are some examples of chemical engineering job titles?

- Biochemical engineer
- Chemical and process engineer
- Refinery engineer
- Chemical development engineer
- Commissioning engineer
- Maintenance engineer
- Process control/automation engineer
- Process safety engineer
- Biomedical engineer
- Research and development engineer
- Sales engineer

Interdisciplinary Engineering

What is interdisciplinary engineering?

Interdisciplinary engineering, or IDE, is a type of engineering that combines knowledge and skills from many disciplines. Several engineering disciplines and specialization alternatives do not fit into any of the major engineering disciplines. Thus, the interdisciplinary engineering domain combines two or more engineering disciplines. This encourages students to build unique skill sets and concentrate in areas that aren't always available in standard engineering degree curricula.

Who are the pioneers of interdisciplinary engineering?

- Steve Slaby (1922–2008) was born in Detroit, Michigan. He was an expert in descriptive geometry, engineering graphics, and the political and social impacts of technology on society. He earned his first degree in mechanical engineering from Lawrence Technological University (previously Lawrence Institute of Technology) in 1943. He served in the U.S. Air Force and later earned a degree in economics from Wayne State University in 1950. He became a professor of engineering graphics (Lafayette College) and also studied labor relations at the University of Oslo. He joined the Princeton University faculty in 1952 and served as a professor of civil engineering and operations research and also taught in the Department of African American Studies.

- Edward F. Leonard (1932–) has been a chemical engineering faculty member at Columbia University since 1958. He earned his first degree in chemical engineering from Massachusetts Institute of Technology in 1953, a master's degree in 1955, and a doctorate degree in 1960. He is most famous for his groundbreaking research (1968) in engineering and the design of artificial organs. He is also a professor of biomedical engineering. Leonard has filed many patents, including one for a blood-filtration device that helps to remove or reduce unwanted substances on a patient's bloodstream. His blood-cleansing system helps kidney patients who depend on dialysis to stay alive.
- Walter G. Vincenti (1917–2019) was a professor of aeronautics and astronautics at Stanford University, and his work laid the foundation for supersonic flight and spacecraft reentry. When he was 10 years old, he saw announcements of Charles Lindbergh's first solo flight from New York to Le Bourget Field in Paris. This birthed his fascination with aeronautics. He earned his first engineering degree in mechanical engineering in 1940 from Stanford. Then, he worked with NASA (originally called the National Advisory Committee for Aeronautics) to develop a mathematical model that would change aircraft design and allow for flight at the speed of sound. He also studied history and how engineers practice and learn how to engineer. He cofounded a new discipline called science, technology, and society (originally values, technology, and society) in 1971. He is considered by many people to be a technology historian.

What are some multidisciplinary engineering concentrations?

- Acoustical engineering
- Educational engineering
- Engineering management
- General engineering
- Humanitarian engineering
- Lighting engineering
- Nanoengineering
- Theater engineering
- Visual design engineering

DID YOU KNOW?

What is the difference between interdisciplinary and multidisciplinary engineering?

Interdisciplinary engineering involves the collaboration of two or more academic fields, such as engineering and medicine, whereas multidisciplinary engineering involves the collaboration of several disciplinary areas.

What is interdisciplinary research?

As per the report of the National Academies of Sciences, Engineering, and Medicine, "Interdisciplinary research is a mode of research by teams or individuals that integrates information, data, techniques, tools, perspectives, concepts, and/or theories from two or more disciplines or bodies of specialized knowledge to advance fundamental understanding or to solve problems whose solutions are beyond the scope of a single discipline or area of research practice."

DISCIPLINES

Environmental Engineering

What is the role of environmental engineers?

Environmental engineers deal with environmental problems using concepts of engineering, soil science, biology, and chemistry. They work to enhance recycling, waste disposal, public health, and pollution management in the water and air. They also deal with worldwide concerns that include contaminated water, environmental degradation, and sustainable development.

Who was the first environmental engineer?

Ellen H. Swallow Richards (1842–1911) was the first environmental engineer. She was also the first woman in America to be accepted to a scientific school.

Ellen H. Swallow Richards was a pioneer in sanitary engineering and helped found the discipline of home economics.

She was admitted into Massachusetts Institute of Technology (MIT). She earned her first degree from Vassar College in Poughkeepsie, New York, then earned a second bachelor of science degree from MIT, and finally a master's degree from Vassar College. Richards is most famous for her water-quality studies. She developed a technique to test the quality of water, and this work confirmed that it was highly polluted. Richards's work led to the state of Massachusetts developing water-quality standards. This work pioneered the field of sanitary engineering.

What are some of the basic duties of environmental engineers?

Environmental engineers are typically responsible for creating reports on environmental investigations. They develop initiatives that help to safeguard the environment, such as water reclamation plants or air pollution control systems. They support environmental cleanup initiatives and legal proceedings with technical assistance. In addition, they have the capability to analyze scientific data, perform quality control tests, and inspect industrial and municipal facilities and programs to ensure that environmental requirements are being met.

What are some of the major challenges addressed by environmental engineers?

The major issues that environmental engineers address are to:

- Supply food, water, and electricity in a sustainable way
- Reduce the effects of climate change and prepare for it
- Create a future free of pollution and waste
- Create cities that are efficient, healthy, and resilient
- Promote intelligent and informed decisions

Humanitarian Engineering

What is humanitarian engineering?

Humanitarian engineering is the practice of using engineering to help those in need. For example, it handles many of the world's issues and humanitarian situations such as helping marginalized people and disadvantaged communities. The majority of projects are community driven and cross-disciplinary with an emphasis on finding simple solutions to fundamental needs such as having access to clean water and having adequate heat and shelter.

What does humanitarian engineering look like?

One example of humanitarian engineering is work that is done through the organization Engineers Without Borders (EWB). This organization partners with communities all over the world to meet basic human needs and prepare more engineers to be ready to face future challenges. It was founded by a University of Colorado civil engineering professor named Dr. Bernard Amadei. He visited a community in Belize who had the resources and desire to have a reliable water supply in their community but lacked the technical knowledge to create a solution that would last a long time. Dr. Amadei worked with a team of engineers to develop a prototype of a much needed solution and then brought 14 students to help him build it. This was the start of the organization Engineers Without Borders.

Engineering Management

What is engineering management? What are the duties and responsibilities of an engineering manager?

Engineering management is the application of management techniques to the field of engineering. It is the practice of improving scientific and technological processes.

The duties and responsibilities of an engineering manager are as follows:

- Organize, coordinate, and direct activities in research, design, and production.
- Supervise a range of tasks using engineering skills.
- Coordinate the development and design of machinery, equipment, systems, and processes.
- Oversee and manage production, operations, quality assurance, testing, and maintenance in industries and organizations.
- Develop budgets for projects and programs as well as staffing, training, and equipment requirements.
- Recruit, allocate, and manage the engineers and support staff who work on the projects.
- Communicate with higher levels of management and coordinate their unit's operations with those of other units or organizations.

What are the major issues and limitations of engineering management?

Major engineering management problems are grouped into the following categories:

- Scaling problems: This includes problems with team sizing and hiring issues, which consist of tasks such as interviewing new candidates, onboarding, coaching, and training new engineers.
- Prioritization problems: Discovering a problem and finding a valid solution is hard and requires strong product management thinking.
- Other problems based on communication, team bonding, and diversity.

Theater Engineering

What is theater engineering?

The use of engineering concepts and functional systems in the design and development of safe, creative, and functional sets, equipment, and entertainment-producing technologies is known as theater engineering. It bridges the gap between the humanities (arts), computer science, and the realm of entertainment arts by combining civil, mechanical, and electrical engineering dis-

DID YOU KNOW?

What is an example of a theater engineering project?

The Prague Quadrennial is an international event that showcases the best acts in the categories of performance design, scenography, and theater architecture. A group of college engineering, theater, and dance students traveled from the United States to Prague, Czech Republic, to design and build two exhibits. From the initial designs to project completion, they worked for over 3,500 hours to make the designs a reality.

ciplines. Engineers with this specialization may work in the theater, film/video production, or amusement parks.

What kind of work do theater engineers do?

Theater engineers are responsible for creating aesthetic and creative set designs or equipment for live entertainment. They also design the next generation of amusement park rides and generate entertainment experiences using engineering problem solving.

Nuclear Engineering

What is nuclear engineering?

Nuclear engineering is the area of engineering concerned with the application of nuclear physics concepts, which includes the breaking down of atomic nuclei (fission) or the fusing of atomic nuclei (fusion), as well as other subatomic phenomena. This engineering discipline is concerned with the study and application of nuclear and radiation processes. These processes include the creation and use of radiation and radioactive materials for uses in research, industry, medicine, and national security as well as the release, control, and usage of nuclear energy.

What are some of the major challenges in nuclear engineering?

Nuclear engineering has several important problems, including maintaining power plant safety, safeguarding reactors from natural catastrophes

DID YOU KNOW?

Who is Kelly Lively?

Kelly Lively is a nuclear engineer who works for the Idaho National Laboratory. She leads the Department of Radioisotope Power Systems, and her work on space nuclear systems and technologies supports space missions. After high school, she started her first career pathway in finance working on quality assurance as an HR secretary. Lively set her eyes on one day working as an engineer. She then earned an electrical engineering degree from Idaho State University.

and external intervention, and developing appropriate, long-term waste-management solutions. Other disadvantages of nuclear engineering include long operation-planning time, possibility of weapon proliferation, risk of meltdown, which includes human error and terrorism, and risk of lung cancer during mining.

Visual Design

What is visual design engineering?

Visual design engineers are responsible for providing artistic form and usability to product design. This can include service or software. Their work is strongly associated with ergonomics and aesthetics, but their technical background distinguishes them from industrial or product designers.

What are some tasks performed by visual design engineers?

- Creating aesthetic and efficient devices
- Designing furniture for comfort and aesthetics
- Performing aerodynamic and passenger safety-related automotive exterior designs
- Creating facility designs and displays that create a visual ambiance while increasing productivity.

Ceramics

What is ceramic engineering? What is the role of a ceramic engineer?

Ceramic engineering is concerned with the use of heat to create things from inorganic and nonmetallic materials, and even high-purity chemical solutions are used to decrease temperatures through precipitation processes. Milling, batching, mixing, forming, drying, firing, and assembling are steps in a conventional ceramic process. A ceramic engineer's main job is to utilize heat to build new goods and equipment using a variety of ceramic materials. Glassware, nuclear reactors, and electronics are among the items that ceramic engineers work on. Ceramics are used in various equipment and objects, including gas burner nozzles, biomedical implants, and jet engine turbine blades.

What are some of the challenges of ceramic engineering?

Challenges in ceramic engineering include understanding unusual phenomena in ceramic microstructures, recognizing the phaselike behavior of interfaces, predicting and controlling heterogeneous microstructures with unforeseen capabilities, controlling the properties of oxide electronics, and understanding defects in the vicinity of interfaces. This also includes controlling ceramics far from equilibrium and accelerating the development of new ceramic materials.

What is a famous ceramics company?

Japanese culture is well known for its commitment to excellence, sense of responsibility, and innovation. The Kyocera Corporation (originally known as

The ceramic tiles on U.S. space shuttles protected them from the intense heat of reentry into the atmosphere.

the Kyoto Ceramic Company) is an electronics company that was founded in Japan by Kazuo Inamori. When it was founded in 1959, its primary product was a ceramic insulator that was used for tubes in televisions. As the company expanded, it began developing technology to be used by NASA and in Silicon Valley innovations, and its current core product line includes ceramic semiconductors.

Who Is Kazuo Inamori?

Kazuo Inamori (1932–) was born in Kagoshima, Japan. He earned a bachelor of engineering degree in 1955 from Kagoshima University. After graduation, Inamori accepted a job from Shofu Industries. During this time, he learned how to synthesize ceramics. In 1959, he launched the Kyoto Ceramic Company (now the Kyocera Corporation) and developed synthetic ceramics that were superior to those being developed in Japan and the United States. His company's success has been attributed to his management and personal philosophies.

Fire Engineering

What Is Fire engineering? What do Fire engineers do?

The implementation of science and engineering concepts to protect people, property, and their environments from the damaging and destructive

The founder of Kyocera Corporation (a ceramics manufacturer) and the telecommunications company KDDI, Kazuo Inamori has contributed to advancements in ceramics and solar cell development.

impacts of fire and smoke is called fire protection engineering. It includes engineering that focuses on fire detection, suppression, and mitigation. A fire engineer ensures that buildings are constructed in such a way that the risk of fire or fire spread is minimized. Fire engineers plan and advise on fire safety measures for both new construction and renovations. Their job is to assist in safeguarding people, property, and the environment from the dangers of fire as well as to guarantee that projects adhere to industry standards and legal requirements.

What are some of a fire engineer's duties?

- Detecting and minimizing potential fire hazards
- Making designs, calculations, sketches, and/or diagrams to aid in the prevention of fire or fire spread
- Sprinkler systems, emergency exits, and fire alarms are all examples of fire safety features that may be included in designs
- Advising on the materials to be utilized in the construction of a structure

What are the required skills for fire engineers to have?

- Understanding of fire, material, and structural behavior
- Fire-testing experience
- Computer fire-modeling experience
- Project management experience

Nanotechnology

What does a nanotechnology engineer do?

- Test for contaminants, make powders to improve meals and medications, and look into the smallest DNA fragments
- Develop creative techniques to test for toxins and pollutants in the air, ground, and water to develop new and improved ways of improving the environment
- Invent innovative devices that can solve issues on a molecular level
- Develop methods for storing and manipulating the smallest quantities of DNA or other biological elements

DID YOU KNOW?

What is nanotechnology?

Nanoscience and nanotechnology are the examination and application of extremely small objects, and they can be utilized in chemistry, biology, physics, materials science, and engineering among other areas. Nanotechnology products are comparatively smaller, cheaper, lighter, and more useful and take less energy and fewer raw resources to produce. Nanotechnology, according to the National Nanotechnology Initiative, is defined as the manipulation of matter having at least one dimension ranging between 1–100 nanometers. The capacity to observe and control individual atoms and molecules is the core of nanoscience and nanotechnology. Nanotechnology has a wide range of applications, including industrial, medical, and energy. These include more robust building materials, therapeutic medication delivery, and environmentally sustainable, high-density hydrogen fuel cells.

- Develop smaller, more efficient chips, cards, and even computer parts to create goods with better or similar performance while generating less waste

What are the advantages and disadvantages of nanotechnology?

The advantages of nanotechnology are as follows:

- It has the potential to transform several industries with the introduction of new materials such as nanotubes and nanoparticles.
- Nanotechnology will pave the way for new ways to generate and store energy. For example, it will make solar energy more affordable by lowering the cost of solar panels and related equipment. Thus, fuel and solar cells are examples of energy-efficient products.
- Nanotechnology has the potential to make significant contributions to medicine. For example, it may be used to make surgeries considerably quicker and more efficient. In addition, it can provide improved imaging and diagnostic technology.
- Electronic technologies such as transistors, LED and plasma displays, and quantum computers have all improved as a result of nanotechnology.

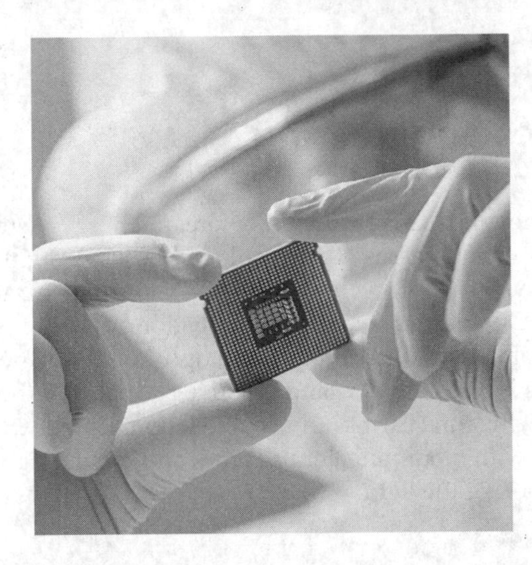

One area that has been greatly improved by nanotechnology is microchips, displays, and quantum computers.

The disadvantages of nanotechnology are as follows:

- Manufacturing technology and employment is greatly affected, as machines often replace skilled labor.
- Automation has a dehumanizing impact; for example, automation requires workers to perform extremely repetitive activities at an ever-increasing rate of production, which might result in physical damage.
- Environmental consequences of nanotechnology include contamination from billions of microscopic, plastic particles in the oceans, waterways, and even our bodies.
- The challenge of keeping up with the newest technologies is generally the most expensive disadvantage of nanotechnology. For example, once a firm commits to a specific technology, the time, effort, and expense of doing so may prohibit the company from upgrading again in a few years as technology advances.
- Other drawbacks include the accessibility of weapons of mass devastation and advanced nuclear weapons and the high cost of research and nanoparticle-based goods.

Who is Dr. Angelique Johnson?

Angelique Johnson (1982–) is the founder and CEO of the start-up company MEMStim, LLC and an adjunct faculty member at the University of Kentucky in the Department of Electrical and Computer Engineering. She earned her first degrees in computer engineering and mathematics from the University of Maryland at Baltimore County. Then, she went to the University

DID YOU KNOW?

What is photonics engineering?

Photonics is the study of how light may be used to create energy, detect information, or transfer data. It encompasses all technological uses of light over the whole spectrum from ultraviolet to visible to near-, mid-, and far-infrared light, including everything from CD players' lasers to the creation of new, energy-efficient light sources to integrated light wave circuits and optical fibers. The primary goal of photonics engineers is to create new and creative products in the fields of health, telecommunications, manufacturing, and construction. Engineers working in photonics are concerned with modifying light sources and regulating the wavelength, intensity, and duration of the light.

of Michigan to earn her master's and doctorate degrees in electrical engineering. She was inspired to pursue a career in engineering by her father, who is a chemical and environmental engineer, and her mother, who is a mathematician and computer programmer. MEMStim, LLC produces implants that treat neurological disorders. She also helps other innovators start technology companies.

Optical and Photonics Engineering

What is optical engineering?

Optical engineering is a branch of science and engineering that deals with the physical phenomena and technology related to the use of light. This also includes creation, transmission, manipulation, and detection. Optical engineers utilize optics to solve issues and develop and create technologies that employ light to accomplish their goals. They develop and manage optical equipment that uses physics and chemistry to harness the characteristics of light. They work in all aspects of optics doing things such as designing lasers, building telescopes, and developing fiber-optic communication systems.

What is lighting engineering?

Lighting engineering is an engineering discipline with a focus on the aesthetics, efficiency, and quality of light and lighting products. Lighting engineers use product design, electrical engineering, civil engineering, and mechanical engineering abilities to create creative lighting arrangement ideas that increase user comfort and safety. Lighting engineers work in interior and architectural design, retail, theater, and construction.

What are some common applications of photonic engineering?

Applications of photonic engineering include spectroscopy, holography, surgery, vision correction, military technology, laser material processing, art diagnostics (which involve X-rays and ultraviolet fluorescence), agriculture, and robotics.

Privacy Engineering

What is privacy engineering?

Privacy engineering is a new discipline of engineering that attempts to develop approaches, tools, and strategies for ensuring that systems guarantee adequate privacy. Privacy engineers make sure that privacy considerations are included into product designs, which may severely affect a company's financial line.

What are some challenges of privacy engineering research?

Some challenges of privacy engineering research are as follows:

- Since privacy engineering is primarily concerned with data minimization, it aims to preserve the user's privacy unconditionally. This threat model isn't always suitable, and it may be too restrictive in practice to produce good results. It can only be met with ineffective protocols that function badly in practice while ignoring the (legitimate) requirements of other stakeholders.
- Other provisions of the General Data Protection Regulation (GDPR), such as subject access rights, transparency, and accountability, are being ignored.
- Current research overlooks the exponential rise of big data and smart algorithms in practice.

Robotics

What is robotics engineering? What does a robotics engineer do?

Robotics engineering combines computer science and engineering. It involves the design, building, operation, and use of robots with an objective to create machines that can assist humans. Robotics engineers are responsible for creating robots and robotic systems that can do tasks that humans are unable or unwilling to perform. Robotics engineers use computer-aided design technologies to build designs that enable them to design down to the circuit level. The robots are then built using computer-aided manufacturing technologies.

What are some challenges of robotics engineering?

Robotics engineering challenges include:

- Developing new materials and fabrication processes to create energy-efficient, autonomous robots.
- Building robots that function more like nature's efficient systems, a battery that matches metabolic conversion, musclelike actuators, self-healing materials, autonomy in any environment, and humanlike sensing, computing, and reasoning.

Robots are being used for more than just industry these days. For example, this police robot can patrol streets in Japan and send video of criminals to its headquarters.

- Robots are inefficient when it comes to conserving energy. Thus, improved power sources are a key concern.
- Communication in robot swarms. Even while acting independently, robots need to communicate with other robots.
- Future robots must be able to function in unmapped and poorly known surroundings.
- Developing AI that has common sense. This can be achieved by combining advanced pattern recognition with model-based reasoning.
- Brain-computer interfaces are costly and inconvenient, and data processing can be difficult. Also, a lengthy period of training, calibration, and learning is required.
- Modeling social dynamics, learning social and moral norms, and developing a robotic theory of mind are all required for creating social robots that truly interact with humans.
- Building reliable medical robots and systems with greater levels of autonomy.
- Violation of robotics ethics is a major concern, which includes problems such as using artificial intelligence in unethical ways and causing the unemployment of workers.

What are the day-to-day activities of a robotics engineer?

- Constructing, configuring, and testing robots
- Creating software systems to operate robotic systems
- Creating automated robotic systems that are utilized to boost production and efficiency
- Assessing prototypes and robotic systems
- Evaluating and approving cost estimates and design calculations
- Assisting with the technological aspects of robotic systems

Telecommunications

What is telecommunications engineering?

Telecommunications engineering is an electrical and computer engineering field that serves to facilitate and improve telecommunications systems. It combines all aspects of electrical engineering such as computer engineering and system engineering. The work covers everything from basic circuit design to wireless network deployment.

What are the duties and responsibilities of a telecommunications engineer?

The duties and responsibilities of a telecommunications engineer are as follows:

- Designing and monitoring the implementation of telecommunications equipment and facilities—for example, electronic switching systems, telephone service facilities, fiber-optic cabling, IP networks, and microwave transmission systems.
- Creating, building, and managing voice and data communications systems—example, fiber-optic, satellite, wired, and unwired connections as well as data encoding, encryption, and compression.
- Identifying users' demands based on input from customers and making required modifications to existing technology to fulfill their needs.
- Programming features, developing systems and networks, and assigning appropriate resources for diverse telecommunication operations.

What are the challenges of telecommunications engineering?

The challenges of telecommunications engineering are as follows:

- The increasing use of the over-the-top (OTT) platform (streaming over the internet instead of using cable and broadcast infrastructure) has created a competitive environment. For example: As compared to traditional short-messaging services, OTT messaging services such as WhatsApp, Messenger, Twitter, and WeChat have shown to be strong alternatives for connecting over the internet. As a result, OTT service providers will continue to deprive telecommunications of significant revenues.
- The way people interact is shifting from voice to Voice Over IP. For example: As more people switch to Wi-Fi on their smartphones, voice traffic has dropped, resulting in a decline in voice revenue. As a result, the average revenue per user has decreased (ARPU). To offset decreasing voice and text revenues, telecom companies now have to develop other revenue streams.

Over-the-top (OTT) apps and other services use the internet to provide streaming and on-demand content.

- Telecommunications companies make fast technological changes. This entails greater expenditures for telecom companies to upgrade infrastructures, purchase new equipment, and conduct new technological development. For example: The evolution of wireless 2G to 3G and 4G.
- Operational support services such as service configuration, order fulfillment, customer care, and invoicing are getting increasingly complicated for telecom operators with millions of customers, a range of new products, and customized solutions. As a result, the expense of conducting these tasks necessitates additional people and technologies, raising the financial burden for administrators and managers.
- Telecommunications firms will need to be transparent and up-front with their customers and clients about how they utilize collected data. They'll also need to make sure they have the proper procedures in place to track the flow of data across the network from start to finish. Any data collected must be properly evaluated and preserved in accordance with industry standards.

What are some recent trends in telecommunications engineering?

Some recent trends in telecommunications engineering are as follows:

- 5G enables telecommunications and other industries to change the way they operate and offer services. Manufacturing, health care, and other industries may all benefit from technological advancements.
- A broad range of services is available in today's telecom environment, all of which require dependable and secure authentication—for example,

the use of smartphones with biometric fingerprint readers are becoming more common. Retailers, financial organizations, the government, and even schools now utilize this technology to verify identities.

- One of the newest telecommunications developments is the integration of artificial intelligence (AI) capabilities to smartphones, allowing them to perform extremely sophisticated activities; for example, telecoms may improve network performance and lower network costs by using data analytics, machine learning, and artificial intelligence. Other examples include the use of speech recognition and indoor navigation.
- Telcoms may generate new income streams by becoming IoT connection service providers and selling machine-to-machine (M2M) devices. In the telecom industry and beyond, the Internet of Things, or IoT, increasingly assists enterprises in meeting urgent demands.
- Telecoms can use microservice architecture to process a large number of requests and transactions as well as to improve their services by adding new operational capabilities.

Petroleum Engineering

What is petroleum engineering?

Petroleum engineering is a branch of engineering that deals with the processes involved in producing hydrocarbons, including crude oil and natural gas. This also includes technological analysis, computer modeling, future production efficiency, and forecasting. Petroleum engineers are also responsible for drilling, producing, processing, and shipping these products as well as dealing with all of the underlying economic and regulatory constraints.

What does a petroleum engineer do?

Petroleum engineering specializations include drilling engineering, production engineering, and reservoir engineering.

- Drilling engineers design and oversee drilling operations in collaboration with geologists and contractors
- Production engineers design procedures and equipment to increase the efficiency of oil and gas extraction
- Reservoir engineers operate as part of a team to establish the best recovery procedures, estimate the number of wells that can be drilled prof-

Oil exploration, extraction, transportation, and refining is a complex and time-consuming process. Petroleum engineers work to improve the process as much as possible.

itably, and forecast future performances using advanced computer models

These engineers have a range of responsibilities, including the following:

- They use technology that can characterize the properties of rocks deep under the surface and detect the types of fluids present inside those rocks
- They develop surface collection and treatment facilities to prepare generated hydrocarbons for transmission to a refinery or pipeline
- They construct and maintain the machinery that raises fluids from underground reservoirs to the surface
- Other tasks of petroleum engineering include pollution cleanup, underground waste disposal, and hydrology

What are some major challenges faced by the oil, gas, and petroleum industries?

- Generating crude oil and refined goods at a reduced cost in order to stay profitable in the business is one challenge. This requires optimizing production systems and environmental utilities on existing operational locations in order to enhance productivity, minimize extraction and refinement costs, and thus compensate for operating expenses.

- The extraction, transportation, and refining processes required to find new oil or gas sources are often expensive and time consuming. This might imply improving equipment efficiency and reliability, lowering dependency on humans through enhanced automation, or creating innovative procedures that minimize the quantity of equipment or personnel entirely.
- In order to get or maintain their operating licenses, companies must re-evaluate extraction, manufacturing, and distribution techniques in order to fulfill increasingly strict environmental criteria. This involves giving assurances and maintaining transparency in terms of environmental management.

OTHER DISCIPLINES

What is acoustical engineering?

Acoustical engineering is a field of engineering that deals with sound and vibration. It involves the use of acoustics, or sound and vibration science, in technology with applications in sound design, analysis, and control. Their main goal is to reduce noise and vibrations that might harm people as well as to enhance the acoustic environment for the general population. A variety of acoustical engineering specialities exist, according to the Acoustical Society of America (ASA), including:

- Architectural acoustics
- Noise control
- Physical acoustics
- Vibration and structural acoustics
- Underwater acoustics

One area of interest to acoustical engineers is noise control, such as getting rid of echoes and other noises in a sound booth at a recording studio.

What is educational engineering? What are the tasks managed by educational engineers?

Educational engineering blends engineering theory and practice with a foundation in education theory and practice to provide a unique skill set. Educational engineers may work in museums, private education networks, not-for-profit education outreach programs, and education-based companies that develop and support engineering learning and educational materials.

Some of the tasks managed by educational engineers include development of appropriate engineering education materials, resources, activities, and teaching/learning artifacts; support of content management systems and educational operations activities; and assessment development, tutoring, and proctoring.

What is the role of an audio engineer?

An audio engineer assists in the creation of a recording or live performance by balancing and modifying sound sources through equalization, dynamics processing, and audio effects as well as mixing, reproduction, and sound reinforcement. Some common job titles for audio engineers are monitor sound engineer, systems engineer, studio sound engineer, research and development audio engineer, wireless microphone engineer, and game audio designer engineer.

What is premedical engineering?

Students pursuing a degree in premedical engineering are preparing for graduate school and ultimately for jobs as medical doctors. Beyond a conventional premedical program of study, these students can use their engineering experience to comprehend the mechanics and operations of the musculoskeletal system. Coursework in engineering, biology, chemistry, anatomy, and physiology may be included in the curriculum.

What is engineering mechanics?

Engineering mechanics is a field concerned with the application of mathematical, scientific, and engineering concepts to solve mechanical problems. Engineering mechanics studies statics, dynamics, material strength, elasticity, viscoelasticity, and fluid dynamics to investigate anything from tiny particles to the tallest structures and analyzes these external influences. Engineering me-

DID YOU KNOW?

What is engineering mathematics?

Engineering mathematics is a field of applied mathematics that deals with mathematical concepts and procedures that are utilized in engineering and industry. It is the technique of using mathematics to solve complicated, real-world issues by integrating mathematical theory, practical engineering, and scientific computers in order to solve today's technological problems.

chanics is used to generate novel concepts and theories, uncover and understand phenomena, and build experimental and coexperimental methods as a bridge between theory and practice.

What does engineering science mean?

Engineering science is the branch of science that examines the physical and mathematical foundations of engineering and machine technology. It refers to a variety of scientific ideas and mathematics that govern engineering. It brings together engineering, biology, chemistry, math, and physics.

What is military engineering?

Military engineering deals with planning and building military constructions as well as maintaining military transportation and communications systems. Military engineers are also in charge of the logistics that support military tactics. Military engineers are in charge of constructing, fortifying, demolishing, surveying, and mapping. Airfields, depots, port facilities, and hospitals are all built by military engineers.

What is health and safety engineering? What is the role of a health and safety engineer?

Health and safety engineering is a field in which engineering, health, and safety expertises are combined to establish processes and design systems that safeguard people from sickness and injury as well as property from harm. It

What is security engineering?

The process of embedding security controls into an information system such that they become an inherent component of the system's operating capabilities is known as security engineering. It focuses on the tools, processes, and techniques needed to design, develop, and test entire systems as well as the tools, processes, and methods needed to modify existing systems as their environment changes.

DID YOU KNOW?

ensures that chemicals, machinery, software, furniture, and other items do not endanger people or cause property damage.

Engineers who work in the field of health and safety usually perform the following activities:

- Regulate the implementation of current health and safety rules and laws and maintain industrial procedures
- Examine new machinery and equipment designs and specifications to ensure that they comply with safety regulations
- Investigate industrial accidents and injuries to find out what caused them and if they might have been avoided or prevented in the future and evaluate product safety
- Inspect facilities, machinery, and safety equipment to evaluate and eliminate risks
- Examine the efficacy of different industrial control techniques
- Ensure that structures or commodities comply with health and safety requirements
- Install or supervise the implementation of safety measures on equipment
- Examine employee safety initiatives and provide suggestions for improvements

What is mining and geological engineering? What are the responsibilities of a mining and geological engineer?

A mining and geological engineer organizes mines for the secure and reliable extraction of minerals (such as coal and metals) for manufacturing and

utility purposes. Mining engineers are employed mostly in remote mining operations, but some also work in sand and gravel operations in larger cities.

The responsibilities of a mining and geological engineer are as follows:

- Designing and managing both open-pit and underground mines, including building mine shafts and tunnels and finding new ways to deliver materials to processing units
- Bringing concrete suggestions to deal with land reclamation, water and air pollution, and environmental sustainability
- Developing technical reports for miners, engineers, and executives and monitoring the operations' productivity

What is agricultural engineering? What is the role of agricultural engineers?

Agricultural engineering is the analysis and implementation of engineering science and design principles for agricultural purposes. This incorporates the concepts of mechanical, civil, electrical, food science, environmental, software, and chemical engineering to enhance the productivity of farms and agricultural business enterprises while also maintaining the sustainability of natural and renewable resources. Agricultural engineers work to address challenges in agriculture related to power sources, machinery efficiency, infrastructure and facility utilization, pollution and environmental concerns, and the storage and processing of agricultural goods.

The duties of agricultural engineers include:

- Designing instruments, techniques, or infrastructures using computer software
- Regulating environmental factors that influence animal or crop output
- Supervising the development of production methods

What is microelectronic engineering?

Microelectronics is an electronic branch concerned with the research, design, and production of very small electronic designs and components. Microelectronic components may be found in computers, tablets, mobile phones, and automobiles. Thus, a microelectronics engineer's main responsibility is to create microchips for a variety of uses, build prototypes for new designs, conduct specialized testing, and collect and evaluate data on new model performance.

DID YOU KNOW?

What is sustainable engineering?

The process of using resources in a way that does not harm the environment or deplete resources for future generations is known as sustainable engineering. It is the practice of developing or managing systems in such a manner that they use energy and resources in a way that is environmentally friendly.

What is paper engineering?

Paper engineering is a field of engineering concerned with the use of physical and biological sciences, as well as mathematics, for the conversion of raw resources into useful paper goods and coproducts. Its aim is on converting raw timber and bark into usable writing, packaging, and other items.

What is geomatics engineering?

The term "geomatics" is described as "the gathering, distribution, storage, analysis, processing, and display of geographic data or geographic information." Thus, geomatics engineering is the management of global infrastructures through the gathering, measurement, monitoring, and archiving of geospatial data.

The specializations in geomatics engineering include:

- Geodetics engineering
- Navigation and positioning engineering
- Geographic information systems (GIS) engineering

What is a geographic information system (GIS)?

A geographic information system (GIS) is a system for creating, managing, analyzing, and mapping various types of data. GIS relates data to a map by combining location data with several forms of descriptive data. Thus, it aims to gather and analyze spatial and geographic data.

What is systems engineering?

Systems engineering is a discipline of engineering and engineering management concerned with the design, integration, and administration of complex systems throughout their life cycles. It uses systems principles and ideas, as well as scientific, technological, and managerial approaches, to enable the effective realization, use, and retirement of engineered systems where systems can be electrical, mechanical, chemical, or biological. In industries such as software, transportation, product development, and manufacturing, systems engineers manage all elements of a project or system. Their role is to design a system that builds a product from start to finish.

CAREERS

Who should choose a career in interdisciplinary engineering?

Interdisciplinary engineering courses are for those who wish to learn about engineering but do not intend to practice it. The program provides a lot of options and allows you to create your own study plan to achieve the educational goals that require you to work at the intersection of engineering and other disciplines.

What types of jobs do interdisciplinary engineers do?

Interdisciplinary engineers can work in the fields of automobiles, aeronautics, power, industrial robotics, medical science, or information technology.

What are some career options in interdisciplinary engineering?

Interdisciplinary engineers work as accountants, computer system managers, marketing managers, CEOs, physicians, dentists, lawyers, pilots, professors, and racecar drivers. In addition, they take roles in research and development, administration, sales, teaching, medicine, and law.

DID YOU KNOW?

In what ways does the automobile sector employ interdisciplinary engineers?

The automobile industry in the 1950s was primarily a place for mechanical engineers to build and maintain IC engines and their related parts. Later on, during the 1990s, changes in technology led nearly all automobiles to have safety measures driven by embedded systems. This was a beginning for electronic engineers to enter the automobile industry. Furthermore, when carmakers investigated the prospect of self-driving automobiles, graduates from different engineering domains started to work in the same sector.

What is the need for interdisciplinary engineering?

Interdisciplinary engineering's primary purpose is that certain engineering tasks need information that is outside the limits (coursework/curriculum) of an engineering degree. For example, to build medical equipment, an engineer must have a solid grasp of anatomy, physiology, biology, and other related topics.

What are the benefits of working in the field of interdisciplinary engineering?

- Program flexibility
- Examining a project using knowledge from disciplines other than fundamental engineering
- Greater productivity and stability
- More consistent work and better income

What are some career pathways to interdisciplinary engineering?

Some universities in the United States, such as Purdue University, offer undergraduate degrees in interdisciplinary engineering. People who take this ca-

reer pathway will use their college years to study the multiple disciplines that they are interested in bringing together to solve a problem. Some people who have careers in an interdisciplinary engineering field have a traditional engineering degree and then follow a career path that is more interdisciplinary in nature.

What are some engineering degrees and their interdisciplinary applications?

- Mechanical engineering: This domain entails an understanding of mass and energy conservation as well as the fundamentals of creating and maintaining machinery, industrial equipment, heating and cooling systems, automobiles, and airplanes. Interdisciplinary applications include the transportation, marine, automobile, and aerospace industries.
- Civil engineering: This domain is concerned with the design and development of buildings, bridges, and highways. Interdisciplinary applications include the environmental and transportation industries.
- Environmental engineering: This domain creates sustainable ecosystems and alternative-energy sources, which necessitates a thorough understanding of both harmful and beneficial chemicals. Interdisciplinary applications include the petrochemical, chemical, and biomedical industries.
- Biomedical engineering: Interdisciplinary applications for this domain mainly consist of the chemical, instrumentation, and pharmaceutical industries.
- Electrical engineering: This domain entails an understanding of how to construct an intelligent power system, communication networks, and chip manufacturing. Interdisciplinary applications include the fields of electronics, communications, computers, and integrated circuits.
- Energy and power engineering: This domain includes learning how to design and build sustainable energy systems that are beneficial to the environment. Interdisciplinary applications include the electrical and environmental engineering industries.

Engineering Pathways

PREPARING FOR COLLEGE

How does someone prepare for getting a degree in engineering?

Here are several steps to help you with preparing for college:

- Research and learn about the different disciplines and fields in engineering
- Learn about the type of work and activities conducted in the area that interests you
- Check labor statistics to find out about the salary and job growth forecast
- Research universities that offer that engineering major
- Look at both public and private universities, keeping in mind how competitive they are
- Check if the programs are ABET-accredited
- Review the admissions requirements and application deadlines

- Check for any required prerequisite courses in the specific department for your major
- Check if the program offers experiential or internship opportunities

What are some major challenges facing science, technology, engineering, and mathematics (STEM) education?

Some major challenges facing STEM education are as follows:

- STEM occupations require proficiency in cognitive skills, critical thinking, and the ability to adapt.
- A belief gap and misconception exists that STEM careers are difficult to pursue.
- A larger percentage of jobs demand higher-education qualifications and skills, whereas just 35 percent of the population in the United States holds a university degree.
- Economic growth centers are usually found in locations away from qualified job-seeking candidates.
- A demographic difference exists based on gender, ethnicity, and income in STEM employment.

COMMUNITY COLLEGE STUDENTS INTERESTED IN ENGINEERING PROGRAMS

Can someone interested in engineering get started in community college?

Community colleges offer two-year degrees, either the Associate of Science (AS) degree or the Associate of Applied Science (AAS) degree. The AS degree is a preengineering degree that prepares students for the prerequisites for a bachelor's degree. The AS degree is intended for students who plan to transfer to an engineering program at a four-year university. The AS degree requires college-level and advanced math and science courses. The specific courses depend on the intended engineering field and the university where you plan to transfer.

The applied degrees typically have engineering technology in their names. Students may also choose to seek certification as professional engineering technicians. The AAS degree is a job-ready credential that prepares students to enter the field as engineering technicians. Students may find that several four-year institutions offer a bachelor's degree in engineering technology and may accept the AAS degree as a transfer credential. The AAS degree typically requires basic math, little to no science courses, and a few general-education core classes.

How can girls become more interested in engineering?

Statistics show that girls are less likely to choose engineering as a career choice. As a result, a national interest has occurred in getting more girls into engineering. The following are a few common recommendations to help girls become more interested in engineering:

- Math and science are essential to engineering. Therefore, early interest and confidence in these subjects are critical for girls.
- Encourage them to ask for help, stay positive, and not be embarrassed or feel defeated if they are experiencing challenges in math and science courses.
- By exposing girls to engineers and engineering experiences, they make tangible connections to the field, building their interest in engineering careers.
- Show girls that engineering can be fun and provide opportunities for them to get involved in activities outside of the classroom.

According to the U.S. Census Bureau, in 2021 women still made up only 17 percent of the engineering workforce in the United States. A concerted effort is being made by the federal government to improve that figure.

DID YOU KNOW?

What courses should I take in high school?

In the application review process for college, your high school curriculum plays a significant role. Taking the high school courses required for graduation by the state may differ slightly from the courses required if you are college bound. For example, the number of years in some subjects and some math and science courses vary. You should challenge yourself by taking rigorous courses such as AP, honors, and IB when available. This will make your application more competitive.

- Help them to stay engaged through high school and beyond.
- Provide mentors to share their experiences, offer advice and encouragement, and be someone they can learn from.

How can families support children with engineering interests?

It is important that children who have engineering or engineering-related interests have the support of their parents and family members. Some homes and families have engineers who can serve as in-home role models, and others do not. Sometimes, support looks like knowing how to solve an engineering problem and spending time doing hands-on projects at home or around the community. Parents and family members who do not have engineering experience should not allow that to hinder them from learning through hands-on project time with their children. Here are a few other strategies for encouraging and supporting children with engineering interests:

- Provide access to engineering toys, games, workshops, and clubs that allow children to engage in building, creating, and designing.
- Discuss your career or hobbies with your child, teach them about what you do, and help them make potential connections between your career and engineering.
- Affirm your child's engineering-related career interest and support their interest by finding role models in the community through professional engineering organizations.
- Establish a tinkering culture around your home whereby when some items are broken, you and your child try to fix or improve them together.

- Encourage your child to spend some time using their imagination and practicing creativity with various objects around the house to build, create, design, and explore.
- Inquire about their STEM classwork and homework. Ask them to solve a problem in front of you or with you.

SPECIFIC HIGH SCHOOL COURSES RECOMMENDED FOR COLLEGE PREPARATION

What is the basic knowledge required to pursue engineering?

Most engineering disciplines require strong foundations in engineering and mathematics. It is important to take these courses seriously in high school. Some engineering students excel in mathematics and science coursework, and others work really hard to do well. An individual does not have to be a math and science genius to be an engineer, but they should learn study skills that will help them succeed.

What are some essential skills required to pursue engineering?

Some essential skills required to pursue engineering are problem solving, research, creativity, innovation, collaboration, attention to detail, passion, and tenacity.

How do I prepare for college if I want to study engineering?

If you start preparing in ninth grade, it is crucial to take the correct courses. Strong performance in math and science is pivotal. Besides taking the basic graduation requirements, explore electives in STEM-related areas such as computer science, robotics, design, programming, and introduction to engineering courses.

What classes should someone interested in engineering take in high school?

At a minimum, students should have four years of English, four years of math (algebra, geometry, trigonometry, and precalculus or calculus), and one year of chemistry or physics.

What AP courses should I take to prepare for engineering school?

You should consider taking any math and science AP courses such as calculus AB/BC, computer science, physics, chemistry, biology, and statistics. Chemistry, biology, and physics AP courses apply to different fields of engineering.

What kind of math is used in engineering?

Engineering uses different types of math, which may vary by discipline and include calculus, differential equations, probability and statistics, and linear algebra.

What is the role of physics in engineering?

Physics provides foundational knowledge that is used in and relevant to most engineering disciplines.

Other than the SAT and ACT, what tests should I take if I want to study engineering?

The SAT and ACT are college entrance tests required by many institutions of higher education. Other tests you may consider are SAT subject tests and AP exams. AP exams are available to all students who feel that they know the subject area. It is not restricted to students who took AP courses in high school. AP exams, if a high enough score is received, can earn you college credits.

DID YOU KNOW?

What types of extracurricular activities should I consider?

Do not limit your activities to those focused on engineering. You can gain transferable skills that you will need in the engineering field from your participation in other activities such as playing a sport, leading a club, participating in a community service activity, or tutoring. In addition, you can find extracurricular activities at your high school as well as in the community.

What can I do during the summer to prepare?

You can find local and national organizations offering summer camps and engineering-related activities through summer programs. Hack-a-thons; coding, math, and science programs; and robotics and rocket camps are just a few options. Some of these activities are offered year-round. Camps and summer programs may require a registration fee, and some may offer scholarships. Often students who are pursuing degrees in engineering work summer engineering-related internships to gain and reinforce skills and knowledge.

Are engineering programs accredited? Why is this important?

Accreditation Board for Engineering and Technology, Inc. (ABET) accreditation provides students with the assurance that a program provides students with the education that not only prepares students for work in the field but also meets the standards set by the field.

What courses do I need to take in college in order to qualify for a specific engineering program?

The courses depend on the area or discipline of engineering you plan to pursue. For example, some colleges require students to take general education

core courses before starting the courses required for the major. Additionally, most engineering majors require math, science, and an introduction to engineering courses. The first year is comprised of introductory courses, which include courses that teach the fundamental skills for each discipline. Differential equations, introduction to engineering, and a programming or statistics course may be common across all disciplines in engineering. Students also take core courses such as English, communications, and social science courses. In the second year, students take foundation courses that build on introductory material and prerequisite courses to prepare for more advanced courses in the third year. In the third and fourth year, students take advanced courses in their area of concentration. Students also consider internships and research opportunities.

Do you need a bachelor's degree to become an engineer?

Most companies require a bachelor's degree to work in the engineering field. However, you can find some positions that only require an associate's degree or vocational training. Without a bachelor's degree, you can work as an engineer technician or in a technology-related area. For example, you can work in cybersecurity, game design, data analytics, or coding.

What are some various jobs/career options after getting a degree in engineering?

A variety of career and job options exist in engineering and typically have high salaries. The possibilities are growing as new problems arise in the world. Engineers touch almost every industry. As the world becomes more and more dependent on technology, engineering fields continue to grow and expand.

What are the minimum admissions requirements for an engineering school?

Admissions requirements can differ across institutions; however, some basic requirements such as a good GPA, an SAT or ACT exam, advanced math and science courses, and being a well-rounded student are consistent. The minimum GPA and SAT or ACT scores also vary. Some universities are moving away from requiring a minimum GPA and test scores. For many top engineering programs, some research experience may give you a boost.

What is the role of LORs (letters of recommendation) in engineering school admissions?

The letter of recommendation is essential and provides the admissions reviewer with a perspective on you from a source who knows you well. For example, a teacher, school guidance counselor, coach, or employer can speak to your character, strengths, and work ethics and provide personal opinions on your potential and other things that are not apparent only by looking at your GPA and test scores.

What documents are required for engineering school admission?

College admission typically requires the same documents for all degree programs. The following items are most common: university admission application or common application, official high school transcript, school report, letters of recommendation, SAT or ACT scores, essays, and an application fee or fee waiver. Some programs may require supplemental materials. If you are applying for a scholarship, you may be required to submit the FAFSA and the CSS financial aid profile.

What does a graduate-school education in engineering include?

A master's or doctorate degree may be required in some fields of engineering for career advancement. Graduate education teaches advanced material and has a more focused curriculum. All courses are discipline specific and cover material not typically taught at the undergraduate level.

Master's programs are 30 credits and designed to take 1.5 to 2 years studying full-time. At the master's level, you can find programs that are project based and programs that are research based. Completing a master's degree not only increases your knowledge but also increases your opportunities to become a supervisor, program manager, or technical lead.

At the doctoral level, you will see two types of programs, the Doctor of Philosophy (Ph.D.) and the Doctor of Engineering (DEng). Both programs are research based, but the DEng is a shorter program and typically intended for working professionals in research-focused positions. On the other hand, the

Ph.D. is research intensive and provides you with an opportunity to learn and develop your research skills. It prepares you for positions as a faculty member or researcher in the engineering field.

Admission to graduate programs may require the following items (each institution may elect to have different requirements):

Master's programs:

- Completion of a bachelor's degree in a related area
- Official transcripts
- A minimum GPA, typically 3.0
- GRE exam
- TOEFL exam (for non-native English speakers)
- Letters of recommendation
- Personal statement

Doctoral programs:

- Completion of a master's degree in a related area
- Completion of a bachelor's degree (some programs only require a bachelor's degree)
- Official transcripts
- A minimum GPA, typically 3.0 (a higher GPA is more competitive)
- GRE exam
- TOEFL exam (for non-native English speakers)
- Letters of recommendation
- Personal statement
- Academic statement

These are common application materials for master's and doctoral programs. Some universities may require additional items, while others may not require some of the items listed. For example, you can find competitive programs that do not require the GRE exam and programs that require a higher minimum GPA. Colleges are moving toward a holistic review of application materials to get a whole picture of each applicant to make an admission decision.

What are some sample first-year undergraduate engineering courses?

- Introduction to Engineering and Engineering Design

Participating in engineering-related clubs can be both educational and a lot of fun!

- Introduction to CAD
- Calculus I/II
- Chemistry I/II and/or Physics I/II
- Communications
- Elective courses

What are some engineering clubs in high school?

Here is a list of some engineering-related clubs you can explore in high school:

- Makerspace
- Rocketry Club
- Robotics Club
- Aviation Club
- Aeronautics Club
- 3D Printing Club
- Coding Club

APPENDIX: ORGANIZATIONS AND ENGINEERING PROGRAMS

Engineering Organizations

- American Indian Science and Engineering Society (www.aises.org)
- American Institute of Chemical Engineering (www.aiche.org)
- American Society of Civil Engineers (www.asce.org)
- American Society of Mechanical Engineers (www.asme.org)
- Association for Computing Machinery (www.acm.org)
- Biomedical Engineering Society (www.bmes.org)
- Center for Minorities and People with Disabilities in Information Technology (cmd-it.org)
- Computing Research Association Committee on Widening Participation in Computing Research (cra.org/cra-wp)
- Institute of Electrical and Electronics Engineering (www.ieee.org)
- Institute of Industrial and Systems Engineering (www.iise.org)
- National Society of Black Engineers (www.nsbe.org)
- Society for Advancement of Chicanos/Hispanics and Native Americans in Science (www.sacnas.org)
- Society of Automotive Engineers International (www.sae.org)
- Society of Hispanic Professional Engineers (www.shpe.org)
- Society of Women Engineers (swe.org)

Engineering Summer Programs and Camps

- Alaska Native Science and Engineering Program (www.ansep.net)
- Bucknell University Engineering Camp (www.bucknell.edu/meet-bucknell/plan-visit/camps-conferences-visit-programs/engineering-camp)

- The College of New Jersey (TCNJ) Engineering in Health and Medicine Summer Camp (bmesummercamp.tcnj.edu)
- Marquette University Opus College of Engineering Engineering Leadership Academies (www.marquette.edu/engineering/k12-outreach/academies.php)
- Max Planck Florida Institute for Neuroscience (MPFI) Machine Learning and AI Summer Camp (https://www.springboard.com/courses/data-science-career-track/)
- Milwaukee School of Engineering (MSOE) Women in Engineering Summer Program (https://www.msoe.edu/about-msoe/k-12-stem-at-msoe/k-12-programs/programs-for-families/women-engineering/)
- MIT Lincoln Laboratory Radar Introduction for Student Engineers (www.ll.mit.edu/outreach/llrise)
- MIT Office of Engineering Outreach Programs (oeop.mit.edu)
- National Aeronautics and Space Administration (NASA) Glenn High School Internship Project (https://www.nasa.gov/centers/glenn/stem/nasa-internship-project-at-glenn)
- Northern Illinois University STEAM Outreach (niusteamcamps.com)
- Science and Engineering Apprenticeship Program (www.nre.navy.mil/education-outreach/k-12-programs/seap)
- Santa Clara University Summer Engineering Seminar (www.uclaextension.edu/engineering/engineering-short-courses)
- Santa Clara University Young Scholars Program (https://www.scu.edu/ysp/)
- Society of Hispanic Professional Engineers (SHPE) Junior Chapters (www.shpe.org/students/junior-chapters)
- Stanford Institutes of Medicine Research (SIMR) Bioengineering Internship (bioebootcamp.sites.stanford.edu):
- Stony Brook University Garcia Center for Polymers at Engineered Interfaces Research Experience for High School Students (https://www.stonybrook.edu/commcms/garcia/summer_program/program_description.php)
- Tulane University STEM Pre-College Summer (summer.tulane.edu/programs/stem)
- University of Florida Research Immersion in Science and Engineering (www.cpet.ufl.edu/students/uf-cpet-summer-programs/research-immersion-in-science-and-engineering)

- University of Notre Dame College of Engineering Introduction to Engineering Program (iep.nd.edu/)
- University of Texas at Arlington College of Engineering Girlgineering (https://www.uta.edu/academics/schools-colleges/engineering/outreach)
- University of Texas at Austin My Introduction to Engineering (cockrell.utexas.edu/eoe/k-12-programs/mite
- University of Texas Medical Branch Summer Biomedical Health Careers Academy (cash4college.net/2021/08/24/summer-biomedical-health-careers-academy/)
- University of Wisconsin at Madison Engineering Summer Program (engineering.wisc.edu/about/inclusion-equity-and-diversity/engineering-summer-program/)

Engineering Programs for Elementary School Students

- The National Society of Black Engineers (seek.nsbe.org)
- The University of Alaska at Anchorage (UAA) Summer Engineering Academies (www.uaa.alaska.edu/academics/college-of-engineering/community/summer - academies)

Engineering Programs for College Students

- Computing Research Association Committee on Widening Participation in Computing Research (CRA-WP) Distributed Research Experiences for Undergraduates (cra.org/cra-wp/dreu/#overview)
- Project Lead the Way (PLTW.org)

FURTHER READING

Introduction to Engineering

"Engineering, a Brief History." LinkEngineering. (n.d.). Retrieved January 8, 2022, from https://www.linkengineering.org/Explore/what-is-engineering/engineering-brief-history.aspx.

Freeman, R. B., and Huang, W. "China's 'Great Leap Forward' in Science and Engineering." NBER. April 13, 2015. https://www.nber.org/papers/w21081.

"Grand Challenges: 14 Grand Challenges for Engineering." Retrieved June 8, 2022, from http://www.engineeringchallenges.org/challenges.aspx.

Grayson, L. P. "A Brief History of Engineering Education in the United States," in *IEEE Transactions on Aerospace and Electronic Systems,* vol. AES-16, no. 3, pp. 373–392, May 1980.

Lu, Yongxiang, ed. *A History of Chinese Science and Technology.* Berlin: Springer Verlag, 2015.

"Wall, Kevin. "Engineering: Issues, Challenges and Opportunities for Development." UNESCO. 2010.

Mechanical Engineering

NASA. "Beginner's Guide to Aeronautics. Retrieved March 12, 2022, from https://www.grc.nasa.gov/WWW/k-12/airplane/.

"What Is MATLAB?" MATLAB & Simulink. (n.d.). Retrieved February 3, 2022, from https://www.mathworks.com/discovery/what-is-matlab.html.

Manufacturing Engineering

Prokopenko, Joseph. *Productivity Management: A Practical Handbook.* Geneva: International Labour Organisation, 1987.

Virasak, L. N. *Manufacturing Processes 45*. Manufacturing Processes 45. Retrieved June 8, 2022, from https://openoregon.pressbooks.pub/manufacturing-processes45/.

"What Is Supply Chain Management?" IBM. (n.d.). Retrieved December 20, 2021, from https://www.ibm.com/topics/automation.

Civil and Architectural Engineering

"Civil Engineering Basics." MyLearning. Retrieved May 22, 2022. https://www.mylearnings.in/civil-engineering-basics/.

Clavero, Jenny. "Fatigue in Construction: How to Prevent It." n.d. eSUB Inc. American Society of Civil Engineers.

"Eddystone Lighthouse." American Society of Civil Engineers. Retrieved January 10, 2022. https://www.asce.org/project/eddystone-lighthouse/.

"History of Civil Engineering." thecivilengg.com. Retrieved May 11, 2022. http://www.thecivilengg.com/History.php.

Engineering Design Process and Innovation

Brown, Tim. *Change by Design: How Design Thinking Transforms Organizations and Inspires Innovation*. Harper Business, 2019.

"Design Heuristics." 2020. http://www.designheuristics.com. IDEO. 2018.

"IDEO Design Thinking." https://designthinking.ideo.com/.

Smithsonian Lemelson Center for the Study of Invention and Innovation. 2022. https://invention.si.edu/.

Industrial Engineering

"Quality Resources. ASQ." Retrieved June 8, 2022, from https://asq.org/quality-resources.

Taylor, Frederick Winslow. *Scientific Management*. New York: Routledge, 2004.

"Toyota Production System: Vision & Philosophy: Company." Toyota Motor Corporation Official Global Website. Retrieved June 8, 2022, from https://global.toyota/en/company/vision-and-philosophy/production-system/.

Bioengineering and Biomedical Engineering

Nebeker, F. "Golden Accomplishments in Biomedical Engineering." *Engineering in Medicine and Biology Magazine*, IEEE, 21(3), 2002, pp. 17–47.

Computer Engineering, Computer Science, and Software Engineering

"Guide to Programming Languages. Code a New Career." Computer Science.org. (2022, April 20). Retrieved June 8, 2022, from https://www.computerscience.org/resources/computer-programming-languages/#c.

IBM Cloud Education. (n.d.). Networking-a-complete-guide. IBM. Retrieved June 8, 2022, from https://www.ibm.com/cloud/learn/networking-a-complete-guide.

Pearson Education India. (n.d.). *Introduction to Computer Science,* 2nd edition. O'Reilly Online Learning. Retrieved June 8, 2022, from https://www.oreilly.com/library/view/introduction-to-computer/9788131760307/xhtml/chapter008.xhtml.

"Telecom Freelance Marketplace for Businesses & Engineers." Telecom Freelance Marketplace for Businesses & Engineers. (n.d.). Retrieved June 8, 2022, from https://www.fieldengineer.com/blogs.

Aerospace Engineering

National Aeronautics and Space Administration. STEM Engagement. Retrieved May 13, 2022. https://www.nasa.gov/stem.

Smithsonian National Air and Space Museum. https://airandspace.si.edu/.

Chemical Engineering

Flavell-While, Claudia. "Claudia Flavell-While Introduces George E Davis, the Man whose Handbook of Chemical Engineering Defined a Profession." The Chemical Engineer. March 1, 2012. https://www.thechemicalengineer.com/features/cewctw-george-e-davis-meet-the-daddy/.

Himmelblau, David M., and James B. Riggs."A Brief History of Chemical Engineering." In *Basic Principles and Calculations in Chemical Engineering*."Hoboken, NJ: Pearson Education, Informit, 2012.

"Visual Encyclopedia of Chemical Engineering Equipment." College of Engineering Chemical Engineering University of Michigan. Retrieved February 20, 2022. https://encyclopedia.che.engin.umich.edu.

Interdisciplinary Engineering

"Profiles."TryEngineering.org. (n.d.). Retrieved May 8, 2022, from https://tryengineering.org/profile/.

"The Women of Idaho National Laboratory's Space Nuclear Team." Energy.gov. (n.d.). Retrieved March 14, 2022, from https://www.energy.gov/diversity/articles/women-idaho-national-laboratorys-space-nuclear-team.

Engineering Pathways

AIChE: The Global Home of Chemical Engineers. https://www.aiche.org.

The Alaska Native Science & Engineering Program. https://www.ansep.net.

American Indian Science and Engineering Society. https://www.aises.org.

American Society of Civil Engineers https://www.asce.org.

The American Society of Mechanical Engineers - ASME. https://www.asme.org.

Association for Computing Machinery. https://www.acm.org.

Biomedical Engineering Society. https://www.bmes.org.

Black in Engineering https://blackinengineering.org/.

Center for Minorities and People with Disabilities in IT. https://cmd-it.org.

European Federation of National Engineering Associations (FEANI). https://www.feani.org.

IEEE—Institute of Electrical and Electronics Engineers https://www.ieee.org.

Institute of Industrial and Systems Engineers. http://www.iise.org.

The National Society of Black Engineers. www.nsbe.org.

SAE International. https://www.sae.org.

Society of Hispanic Professional Engineers. https://www.shpe.org.

Society for Advancement of Chicanos/Hispanics and Native Americans in Science. https://www.sacnas.org.

Society of Women in Engineering. https://swe.org.

INDEX

NOTE: (ILL.) INDICATES PHOTOS AND ILLUSTRATIONS

D

E

F

G

J

K

L

M

N

O

U

V

W

X–Z